WITHDRAWN

Volume III
The Foundations of Ethics
and Its Relationship
to Science

Morals
Science and
Sociality

Edited by
H. Tristram Engelhardt, Jr.
and Daniel Callahan

THE HASTINGS CENTER
Institute of Society, Ethics and the Life Sciences

Institute of Society, Ethics and the Life Sciences
360 Broadway
Hastings-on-Hudson, New York 10706

Library of Congress Cataloging and Publication Data

Main entry under title:

Morals, science, and sociality.

   (The Foundations of ethics and its relationship to
   science ; v. 3)
   Includes bibliographical references and index.
   1. Science and ethics—Addresses, essays, lectures.
   I. Engelhardt, Hugo Tristram, 1941-   II. Callahan,
   Daniel, 1930-   III. Series.
BJ57.M67   174'.9'5   78-14481
ISBN 0-916558-03-7

Printed in the United States of America

Contents

iii

Contributors

RICHARD D. ALEXANDER is Professor of Biology in the Division of Biological Sciences and Curator of Insects in the Museum of Zoology of The University of Michigan. Since 1971 he has published about fifteen articles on the significance of evolutionary biology for human understanding of humans.

DANIEL CALLAHAN is Director of the Institute of Society, Ethics and the Life Sciences. He received his B.A. at Yale and his Ph.D. in Philosophy at Harvard. He is the author, most recently, of *The Tyranny of Survival* and *Abortion: Law, Choice and Morality*. He is a member of the Institute of Medicine, National Academy of Sciences.

ERIC J. CASSELL, M.D., is Clinical Professor of Public Health at Cornell University Medical College and a Diplomate of Internal Medicine in private practice. For the past several years Dr. Cassell has been doing research and writing on the underlying bases of medical practice and the relationship between doctors, patients, and disease. He is the author of *The Healer's Art*.

GERALD DWORKIN is Professor of Philosophy at the University of Illinois at Chicago Circle. His main interests are in moral, political, and legal philosophy. He is the co-editor of *The I.Q. Controversy* and *Markets and Morals*.

H. TRISTRAM ENGELHARDT, JR., Ph.D., M.D., is Rosemary Kennedy Professor of the Philosophy of Medicine, Kennedy Institute, Center for Bioethics, Georgetown University. He is the author of *Mind-Body: A Categorial Relation* and co-editor of *Evaluation and Explanation in the Biomedical Sciences*, and *Philosophical Dimensions of the Neuro-Medical Sciences*.

LOREN GRAHAM has recently joined the faculty at the Massachusetts Institute of Technology as Professor of the Social and Historical Study of Science. Until 1978 he was a professor of history at Columbia University. His publications include *The Soviet Academy of Sciences and the Communist Party* (1967) and *Science and Philosophy in the Soviet Union* (1972). He has been a member of the Institute for Advanced Study and has held Guggenheim and Rockefeller Foundation Fellowships. He is currently writing a book on the relationship of science to sociopolitical values.

MARJORIE GRENE is Professor of Philosophy at the University of California, Davis. She received her Ph.D. at Radcliffe College, Harvard University. Among her most recent publications are *Sartre, The Understanding of Nature: Essays in Philosophy of Biology*, and *Philosophy In and Out of Europe*. She is chairman of the program committee, American Philosophical Association, Pacific Division.

ALASDAIR MACINTYRE is University Professor of Philosophy and Political Science at Boston University. He is the author of *A Short History of Ethics* and *Against the Self-Images of the Age: Essays in Ideology and Philosophy*.

STEVEN MARCUS is George Delacorte Professor of the Humanities at Columbia University and Director of Planning, National Humanities Center. He has held Guggenheim and Fulbright Fellowships and is a Fellow of the American Academy of Arts and Sciences. Among his books are *The Other Victorians* and *Engels, Manchester and the Working Class*.

JOSEPH MARGOLIS is Professor of Philosophy at Temple University and Editor, PHILOSOPHICAL MONOGRAPHS. His most recent books include *Persons and Minds* (1977) and revised editions of *An Introduction to Philosophical Inquiry* (1978). *Art and Philosophy* is forthcoming.

PAUL RAMSEY is Harrington Spear Paine Professor of Religion, Princeton University, and author of *The Patient as Person* and *Ethics at the Edges of Life: Medical and Legal Intersections*.

KENNETH F. SCHAFFNER is Professor of History and Philosophy of Science and of Philosophy, and Co-Chairman of the Program for Human Values in Health Care at the University of Pittsburgh. He received his Ph.D. in philosophy at Columbia. He is Editor-in-Chief of *Philosophy of Science*, the author of *Nineteenth-century Aether Theories*, and is presently completing a book on *Discovery and Explanation in the Biomedical Sciences*.

ROBERT C. SOLOMON is Professor of Philosophy at the University of Texas at Austin, having taught at Princeton University, the University of Michigan, UCLA, and C.U.N.Y. He is the author of *From Rationalism to Existentialism* (1972) and the editor of *Nietzsche* (1973) and *Existentialism* (1974). His latest book is *The Passions* (1976).

GUNTHER S. STENT is Professor of Molecular Biology at the University of California, Berkeley. His current scientific research interest is the function and development of the nervous system. He is the author of *The Coming of the Golden Age* and of *Molecular Genetics: An Introductory Narrative*.

STEPHEN TOULMIN is a member of the Committee on Social Thought at the University of Chicago. He has written extensively about philosophical and historical aspects of science, as well as about ethics and practical reasoning. He is at present in the course of completing a large-scale work on *Human Understanding*.

GREGORY VLASTOS is Stuart Professor of Philosophy, Emeritus, Princeton University and Mills Visiting Professor of Philosophy, University of California, Berkeley. He is the author of *Platonic Studies* (1973) and *Plato's Universe* (1975).

Preface

WHEN THE HASTINGS CENTER, Institute of Society, Ethics and the Life Sciences, was founded in 1969, the word "ethics" conjured up for most academics and intellectuals something at best mysterious and at worst obscurantist. The dominant positivism of the forties and fifties together with an attendant cultural relativism put the whole subject of ethics and morality in a dark shade. When The Hastings Center proposed to address itself to certain critical ethical issues emerging in biology and medicine, few were prone to deny the reality of the issues or the knotty dilemmas they posed: the care of dying patients, psychosurgery, genetic engineering, behavior control, excessive population growth, and the allocation of scarce medical resources.

What many were prepared to deny, however, was that ethics could be a fit subject for intelligent, even illuminating, discourse. Was it not the case that philosophers have wrangled interminably and to little avail over the meaning and implications of words such as "right," "good," "duty," and "obligation"? Was it not also the case that human behavior is shaped far more by family, state, and culture than by anything moralists had ever said? And in any event, if there was to be any knowledge of ethics, the sciences were far more likely to produce insight than would philosophy or theology.

Fortunately, the very urgency of the issues we set out to examine helped to overcome many of the doubts. Ethics was at least worth a new gamble, if only because no one could offer a better way of proceeding. One way or another, decisions have to be made, and ethics is there whether one likes it or not. The ensuing years saw a rebirth of interest in normative ethics within departments of philosophy, a new respect for moral reasoning within biology and medicine, and the emergence of a fresh literature in what has come to be called "applied ethics." The term "applied ethics" is an odd one. If the whole point of reflection on ethical matters is not application, what else could its point be? Despite the validity of descriptive ethics, on the one hand, and of metaethics, on the other, something had been missing in contemporary ethics—a systematic attempt to develop modes of practical moral reasoning sufficient to allow moral decisions to be made. That is the old-fashioned field of normative ethics, and the term "applied ethics" is simply another way of characterizing that critical domain.

Much of the work of The Hastings Center has focused on moral decision making in concrete cases, ranging from decisions on the control of recombinant DNA research through the often nasty choices faced by physicians in terminating treatment on a dying patient. We believe this has been a valid and helpful emphasis. Nonetheless, as time went on it became increasingly clear that applied ethics requires the support, indeed the firm backbone, of more general theory. In particular, it is evident that in our time science in both its serious and popular manifestations has had an enormous impact upon ethical thinking. But to what extent is that impact justified? Does science have something of enduring validity to contribute to the way we think about the moral life and the ends of human life? In turn, does ethics have something of validity to say to science?

Those are troubling and muddy questions, but they have to be taken on. The point of our project on "the foundations of ethics and its relationship to science" has been to foster that engagement. This is the third volume of a planned four-volume series. The first volume, *Science, Ethics and Medicine* (1976), while focusing on the general theme, gave special attention to medicine. The second volume, though continuing and extending the general discussion, had as its special focus the relationship between ethics and religious belief. That volume was entitled *Knowledge, Value and Belief* (1977). The general discussion has gone forward in this the third volume but with a subemphasis on the impact of scientific thinking on ethics, both theoretical and popular.

While the theme of the series has been the relationship between science and ethics, an important practical problem has always presented itself. Who is best suited to talk about that relationship? The answer seemed obvious enough: those who know something about ethics and those who know something about science. But there is very little in the way of recent tradition to help carry forward that kind of a dialogue. For the most part, C.P. Snow was right when he complained of the "two cultures" in western society. Our project has been forced to work across the boundaries of those two cultures, not just at the level of theory but at the sometimes even more difficult level of ordinary human relationships. In preparing these volumes, nothing less than a full-scale interdisciplinary inquiry has been needed. It has been both difficult and a pleasure. Difficult because philosophers, theologians, scientists, and historians do not talk the same language in their professional life, but a pleasure because to some measure we have learned over the years to understand and appreciate each other's ways of looking at the world.

This project has been supported by a grant from the National Endowment for the Humanities, that remarkable organization which has helped to breathe new life into humanistic inquiry. As we have done before, we particularly want to express our thanks to Richard Hedrich of the Endowment for his assistance and support. As with any good foundation officer, he realizes that the best form of assistance is that of the firm but genial nag, constantly probing for the weak spots in our work and persistently pointing out the gap between the high rhetoric of a proposal and the day-to-day reality of a solid research project.

Daniel Callahan
H. Tristram Engelhardt, Jr.

Part I

The Moral Roots of Science

Morals, Science and Sociality

H. Tristram Engelhardt, Jr.

El universo (que otros llaman la Biblioteca) se compone de un
número indefinido, y tal vez infinito, de galerías hexagonales, con
vastos pozos de ventilación en el medio, cercados por barandas
bajísimas . . . *La Biblioteca es ilimitada y periódica.* Si un eterno
viajéro la atravesara en cualquier dirección, comprobaría al cabo
de los siglos que los mismos volúmenes se repiten en el mismo
desorden (que, repetido, sería un orden: el Orden). Jorge Luis
Borges, "La Biblioteca de Babel."[1]

We approach reality as if we could decipher from it an ingredient meaning beyond that which we as knowers and doers bring
to it. Like Borges's librarians in the library of the universe, we
seek ways to decipher its writings. But, as Borges's story suggests, the only way we can encounter reality is through our needs
for order and rationality. We must impose order upon reality in
the process of discovering the order within it.[2] This is as true of
the world of ethics as it is of the world of science. A minimum
level of coherence is a necessary condition for the possibility of
the domain of blame and praise and moral goods, as it is for the
domain of scientific explanation, prediction, and control of reality.[3]

There may, as a result, be greater similarities than one might
otherwise suspect between ethics (and enterprises of evaluation
generally) and the sciences. Insofar as we through knowing and

1

doing in the world construct its meanings, there should be features common to the domain of evaluation and explanation. Both reflect human purposes. Both are done by reasoning creatures. Or, less subjectively, if we recognize reality as consanguineous with our basis categories of knowing and doing, we should expect to find, and will undoubtedly find, similarities between the foundations of ethics and the foundations of the sciences.[4] Systematically doing, acting, that is conducting ourselves morally, may not be so different from deporting ourselves as systematic knowers. In choosing among ethics and among scientific accounts, we in both cases choose those accounts with greater explanatory compass, fewer internal conundrums, greater congruence with the facts, and greater success in signaling novel or future issues. Since science and ethics are both forms of reasoning they should both show common marks of rationality.

Besides common roots in rationality, science and projects of evaluation overlap through borrowing ideas and borrowing methods. Thus, science is often evaluative and ethics is often dependent upon scientific data. For example, in knowing the world we often impose categories that classify elements of reality with respect to their satisfying or failing to satisfy goals *we* have in mind (consider the concepts of health and disease, or overpopulation),[5] so that in the process of knowing the world we often evaluate it. Many concepts have, in short, both normative and descriptive dimensions. Many terms also have a social force, casting individuals into particular social roles and molding social structures (e.g., "X social structure should be supported because it is the product of the survival of the fittest").[6] Moreover, any concrete specification of moral duties will also require reference to particular empirical facts about humans. Concrete views of ethics, accounts of human values, presuppose concrete, empirical accounts of human nature and the nature of reality. As a result, views of the human condition have an impact upon what we take, at least popularly, to be proper human goals and expectations. One need here think only of the impact that Charles Darwin and Sigmund Freud had upon popular considerations of mores. Furthermore, insofar as ought implies can, and descriptions of the world indicate the relevance of particular moral ideals (are some individuals, as Aristotle argued, natural slaves?),[7] empirical data about ourselves will influence accounts of ethics. And to complete the circle, any passably detailed account of the development of scientific theories will need to examine the appeal of particular genres of scientific theory in different cultural ambients to show how the surrounding ethos affects the sciences that give the data for concrete ethics—points made by Loren Graham and Stephen Toulmin in this volume. The cultural ambient influences the articulation and acceptance of scientific theories (e.g., as shown in accounts of the context of discovery) even if it does not bear directly on the rational justification of scientific theories.

One might think here of the attempt by Ludwik Fleck[8] and those who followed him (e.g., Thomas Kuhn)[9] to develop sociological epistemologies: communities of researchers develop facts in terms of their research community's way of viewing the world, style of thought (*Denkstil*), which can be influenced by numerous cultural factors. Facts then have their being in terms of their standing for a community of researchers. Scientific research is defined within a community of thought (*ein Denkkollektiv*); a "fact must be experienced in the style of the community of thought . . . every fact must be in line with the interests of a community of thought."[10] The result is a view of science, which, many argue, including Alasdair MacIntyre in this volume, erodes an important sense of objectivity—correspondence to the objects of knowledge. Science runs the risk of losing its traditional commitment to explaining reality. In any event, objectivity runs the risk of being fragmented into many somewhat discontinuous domains of reason, with disparate views of reality. If one cannot supply a paradigm of paradigms, a set of transcendental conditions, one may have numerous limited areas of intersubjective agreement, but no common, integrated reality. The same has been argued in ethics. There are many stories, as Burrell and Hauerwas argued in the last volume in this series, but there is no story of the stories—no universal arguments to sustain consensus.[11] One is rather confined, as Burrell, Hauerwas, Kuhn, *et alia* have contended, within particular traditions. If they are correct, then science and ethics are parochial. One is thus faced with issues core to delineating the foundations of ethics and its relationship to science; are there many grammars of knowing and evaluating, but no grammar of grammars? Answering such questions is a condition for understanding the relationship of ethics and science, because those answers will tell us how different areas of human reason and reality are related.

These questions have been addressed in the meetings that produced the volumes in this series, the Foundations of Ethics and its Relationship to Science. These questions arise out of a concern about the unity of reason, about the consanguinity of rational attempts to come to terms with reality. This project has focused on discovering and examining the roots that nourish the diverse domains of ethics and science. Or, to use Fleck's idiom, this project has been examining how communities of thought in ethics function, and how the styles of thought of such communities influence and are influenced by the styles of thought of scientific communities. The lines of influence traced have been both causal (e.g., historical) as well as conceptual (i.e., logical relations between views of rationality, in ethics and in science). The result has been a series of four meetings over three years, leading to the production of papers which have, for the most part, been evenly divided between focusing upon the foundations of ethics considered separately, and areas of overlapping subject matter between ethics and the sciences. A number of themes have begun to emerge with regard to basic conceptual presuppositions in reasoning both in ethics and the sciences. And, beyond considerations of the interplay of the subject matter of ethics and the subject matter of the sciences, questions on a methodological level have arisen as well. These various themes can be summarized as follows:

I. Concepts guide both science and ethics, not simply through their conceptual content, but through their metaphorical force.

II. There are causal explanations available to account for the existence of moral dispositions; science can appropriate ethics, in part, by giving an empirical explanatory account of ethics.

III. There is a recurrent attempt to secure the status of some values by attempting to read ethical norms from the processes of nature (e.g., evolution).

IV. Both science and ethics presuppose value judgements; there is no crisp line to be drawn between an explanation and an evaluation, between value-neutral facts and fact-free values; value theory can thus come to appropriate part of the subject matter of science as its own subject matter.

V. Though both ethics and science may make claims of universal truth or applicability for their axioms, the articulation of both scientific and ethical assertions is time-bound, placed within particular historical matrices. Though reason may presuppose universality as a regulative ideal, in any particular case reasonings are historically and culturally parochial.

As these five themes indicate, the goal has been interdisciplinary. The first volume addressed these issues with a special accent upon medicine, and the second volume gave particular attention to the bearing of religion and theology upon the understanding of ethics and its relationship to science. This third volume continues these interdisciplinary explorations, but with a special consideration of the role of social and psychological factors on the relationship of ethics and the sciences. Here, we in particular hope to indicate both how scientific views are taken to support and/or change mores, and how mores influence the development of science and direct scientists. In short, we will press our examination of the relationship among the enterprises of evaluation and explanation.

Alasdair MacIntyre opens this volume by addressing the nature of objectivity in science and morality. He does this by suggesting the lineaments of a critique of subjectivism and relativism in ethics and in the philosophy of science. In this essay the focus falls upon science and the philosophy of science. MacIntyre provides an argument (in part, following Michael Polanyi)[12] in favor of the roles of consensus, tradition, and authority in the scientific community and against the incommensurability theses of Thomas Kuhn and Paul Feyerabend.[13] Science, so MacIntyre argues, does not fall apart into various communities of researchers with noncomparable views of reality. The ideal of science as the pursuit of an adequate account of reality is not to be lost in a relativism. On the contrary, the scientific community is bound together by an interest in the internal goods of science which include the pursuit of true representations of nature. Science becomes, or rather truly is, "a morality" (MacIntyre, p. 38). That is, "The scientific community is one among the moral communities of mankind and its unity is unintelligible apart from the commitment to realism" (MacIntyre, p. 37). Scientists are thus bound together in one community by reference to this commitment to realism (and, perhaps, other internal goods of science).

MacIntyre thus gives a moral grounding for scientific objectivity and for science. "Objectivity is a moral concept before it is a methodological concept and the activities of natural science turn out to be species of moral activity" (MacIntyre, p. 37). The result is a Kantian redemption of the commitment to realism; the sense of the scientific enterprise requires a commitment to realism as a regulative ideal. "The building of a representation of nature is in the modern world a task analogous to the building of a cathedral in the medieval world or to the founding and construction of a city in the ancient world, tasks which might also turn out to be interminable" (MacIntyre, p. 37). The consequence of this line of argument is to make ethics the foundation of science. MacIntyre's answer to the relationship between the foundation of ethics and science is that one must give "morality primacy over science. For science can only become intelligible to us through its history and the continuities of this history are moral continuities. Our knowledge of nature is, we know, immensely fallible . . . [O]ur knowledge of our knowledge of nature, that is, our knowledge of the history and philosophy of science is, [however], far better founded. Yet its foundation depends in turn on our understanding of the history of human relationship to goods." (MacIntyre, pp. 38-39). To understand science one must first understand its morality. As a result, the truths of morality are better grounded than those of science—a suggestion to which Marjorie Grene takes exception in her commentary.

To this discussion of the moral foundations of science Stephen Toulmin offers a proposal for a moral psychology of scientists and of their engagement in science. Toulmin suggests that we have wanted to view scientists as solitary embodiments of reason, pursuing science as a value-free, cognitive endeavor and as if it were only an end in itself. An adequate account of scientists and of the communities they form would require more. One would want to know, among many things, why certain good (bad) arguments are not accepted in some historical periods but are accepted in other historical periods. "The 'intrinsic power' of novel scientific arguments is matched by an equal 'effective power'—their proper intellectual merits are translated into an operative capacity to persuade—only if they are directly relevant to the currently accepted concepts, theories and problems of the science concerned" (Toulmin, p. 53). One would want as well to know why individuals pursue scientific careers and the nature of the secondary gains that science secures for them. Moreover, Toulmin suggests that this austere view of the scientist disinterestedly pursuing science is not only wrong, but may, as well, be dangerous. "The 'purity' of the scientific life has its own powerful attractions, which easily seem to justify a turning away from ethical issues. So, once we are launched into an examination of the moral psychology of science, in both its individual and collective aspects, we may find that the *vita scientifica* involves ethical difficulties and complexities no less acute than those traditionally associated with the *vita monastica*, which was its prototype and forerunner" (Toulmin, p. 66). A moral psychology of science may thus contribute to an ethics for science and scientists.

In response to these contentions, Gunther Stent indicates that the role of psychological, ethical and political aspects of science have been increasingly recognized, pressing the need for an adequate moral psychology of scientists and presumably of the social psychology of the community of scientists (here containing much more than a sociology of knowledge).

The papers by Steven Marcus and Joseph Margolis continue the examination of the roles of values in the sciences with a focus upon Sigmund Freud and psychoanalysis. In his account of the development of the early Freud and his works, Marcus provides what is tantamount to a moral psychology of Freud. He portrays Freud as an individual committed to the internal goods of science, to use MacIntyre's idiom. "Freud's intellectual commitment and adherence to the idea of science has a profoundly moral component to it" (Marcus, p. 80). Freud not only had a commitment to realism in science but had an even more passionate commitment to the possibility of discovering meaning in the patterns of human conduct. "Freud himself was not religious in either temperament or sympathy. Nevertheless, like the other great modern figures he retained one belief, albeit in secular form, that belongs to that tradition. This is the belief in meaning—a belief of extensive moral and cultural consequence. In the instance of Freud it was a belief that all human thought and behavior had a meaning and meanings—that they were understandable, purposeful, had structure and rose to significance" (Marcus, p. 87). The result was a systematic analysis of human behavior which took account of values and cultural contexts.

Joseph Margolis's analysis is an attempt to vindicate this general enterprise—to show the possibility of a cultural science as undertaken by Freud. Margolis shows why an adequate account of human conduct must recognize emergent qualities—a reductionism will not succeed. In fact, only by reference to cultures and institutions can one specify particular human behaviors as conduct, actions. The kind of causal explanation used by a cultural science such as psychoanalysis must fall under "covering institutions," not covering laws (Margolis, p. 107). On the one hand this defends the possibility of Freud's attempt to create a science of human behavior. On the other hand, *pace* Freud, it means that because persons are culturally emergent entities, there are no naturalistic ways to specify ethical norms. One is left with a "robust relativism" (p. 113) and a need to presume freedom (Margolis, p. 111) as a consequence of having succeeded in defending a scientific study of human conduct. One is, moreover, left with a view of such a science as essentially concerned with values. Thus in moving from MacIntyre's essay to that of Margolis, one moves from an argument that science is grounded in morality to an argument that a dimension of science is essentially concerned with morality (i.e., human mores, unreduced-away values).

Science is also rooted in values in that the acceptance of scientific theories and their consequences turn on how the sciences are interpreted in particular cultural milieus. As a case in point, Loren Graham sketches the debates in Germany and Soviet Russia during the 1920s concerning Mendelian versus Lamarckian interpretations of eugenics. Though the National Socialists finally embraced a Mendelian interpretation, and the Communists a Lamarckian interpretation, Graham shows that in principle the debate could have gone either way, especially in Russia. There were good arguments to show that a Mendelian interpretation was fully compatible with a Marxist-Leninist account of society and history, if not more so than a Lamarckian view. What was decisive were what Graham terms "second-order" links between science and values: 1) contingent political and social situations, 2) prevailing technological capabilities, and 3) the persuasiveness of particular ideological interpretations. Though sciences may be value-free considered abstractly, they are in any actual case immersed in an historical situation. They have, as a result, value implications because of the technologies they can spawn. Graham's case study is important in showing how interpretations of the links between a science and its possible technologies influence the acceptability of that science, and the development of the character of the scientific community. Again, values and the scientific community have important and mutually defining interdependencies.

But in addition to such external implications of scientific theory, Kenneth Schaffner argues that there are values internal to science as well. In principle, and even in the abstract, science cannot be value-free. In addition, sciences may have external value implications apart from the technologies with which they may be associated. One must, in short, look painstakingly at the various levels of relations between sciences and values. One must "direct attention at the valuational assumptions in those social contexts [of the sciences], at alternative assumptions of a valuational sort, at the ethical foundations of such valuational assumptions, at the implications produced by the scientific theories under alternative ethical assumptions, and at the methodological foundations of the adjudication process in which we examine the valuational and scientific assumptions in the light of conditions and consequences." (Schaffner, pp. 154-55) The relationship between science and values is many-layered and profound.

The next three sets of papers and their commentaries in this volume concern the foundations of ethics itself, though with important implications for the sciences. Each of the papers deals with the place of the self (or person or individual) in moral reasoning; in each case the paper raises issues which can in part be illuminated by empirical science. The papers by Gerald

Dworkin and Gregory Vlastos deal with the relationship between the individual and his community with its traditions. Dworkin presents a critique and analysis of the view of the moral agent as an autonomous individual. This inquiry leads to an examination of the extent to which a person's moral principles can be (or should be) his own. As Dworkin shows, there is no way to make sense of the moral life without recognizing individuals within traditions or moral contexts. Each individual does not start the moral life *de novo*. "We are born in a given environment with a given set of biological endowments. We mature more slowly than other animals and are deeply influenced by parents, siblings, peers, culture, class, climate, schools, accident, genes, and the accumulated history of the species. It makes no more sense to suppose we invent the moral law for ourselves." (Dworkin, p. 158) Dworkin thus supports a conception of moral autonomy as critical reflection within the limited authority given by tradition.

In contrast, by providing an analysis of Plato's polis in which basic (extra-political) human rights have no standing, where all rights are instead defined by reference to rights within the polis, Gregory Vlastos gives a critique of a too heavy reliance upon tradition and context in the moral life. Vlastos characterizes the principle of justice and of the origin of rights in the *Republic* as the Principle of Functional Reciprocity. "All members of the polis have equal right to those and only those benefits which are required for the optimal performance of their function in the polis." (Vlastos, p. 178) Rights, perquisites, duties are derived from one's function within the polis, within the community. Though this view led to the expansion of some rights, Plato fails to give any account of human autonomy, or to be concerned about it.[14] As Vlastos indicates in criticism, individuals in the Republic had no moral standing of their own. Vlastos's Plato is thus on the opposite extreme from Dworkin's autonomous individualist who wishes to create all moral principles *de novo*. Through this case Vlastos sketches the need for a balance between a commitment to the polis and the pursuit of individual autonomy. It is worth noting that some of the presumptions regarding the relationship between individuals and their tradition may in part be empirical—that is, socio-psychological and biological. In any event, as the commentary on these papers indicates, they offer reciprocal insights into the balance between individual freedom and individual dependence upon the social context. Each side of the balance is by itself one-sided and abstract.

The analysis of persons given by Eric Cassell internalizes the fragmented world of Burrell and Hauerwas's stories, Kuhn's paradigms, and Fleck's communities of thought. To account for the ways in which persons come into conflict within themselves in making moral decisions, Cassell argues that one must posit the existence of various selves. He gives, as an example, the conflict between a person's adult intellectual self and her child self in making a decision about the treatment of a dying parent. The selves are distinguished in terms of differences in 1) the rules of conduct that constitute them, and/or 2) the beliefs that they hold, and/or 3) the perceptions they have of the world. In successful maturation these selves merge into one person who through experience knows the world and makes himself in the process. But, as Paul Ramsey notes in his commentary, if there is a person who succeeds in making these selves one, how strong is the claim that we have many selves. Is it closer to a weaker claim that Cassell at times appears to be making: that having different selves is like having different masks (personae), different perspectives one can assume? If he means to forward a strong claim, one runs into the problem of apportioning moral responsibility for past actions among the selves. The problems concerning the coherence of external reality are thus encountered with regard to the coherence of the internal reality of persons. Paul Ramsey suggests the general Kantian remedy: the possibility of experience (and of action) requires presuming a unity of both reality and of the experiences (or agent) in reality. But that ploy may be a way of begging the question, how many unities are we? Whether we are at times many selves may indeed, as Cassell argues, be an empirical issue—one that if true would have important implications for the foundations of ethics.

After discussing the foundations of ethics, this volume turns to the origin of ethics: Richard Alexander gives a socio-biological account of the origin of moral rules (e.g., the Ten Commandments) and laws. He does this while clearly disavowing any normative claim for the derivation (Alexander, pp. 277-78). We derive important information concerning ourselves from the fact that rules and laws have developed in order to increase reproductive success, but that does not tell us what we ought to do. Alexander, however, advances a number of interesting explanations of human behavior and of the presence of societal laws. He remarks that though we are similar to other primates with whom we share recent ancestors, and though we are similar to group hunters such as dogs, wolves, and lions, from sharing a similar past relationship with our environment, we are somewhat unique in having evolved through competition and predatory relations with others of our own species. Alexander therefore advances what he terms the "balance-of-power hypothesis." That hypothesis, along with other principles (e.g., "Happiness is a means to reproduction," Alexander, p. 265), allows him to forward a number of explanations for diverse facts about human behavior. He addresses such phenomenon as the greater incidence of law-breaking among men vis-a-vis women, and the fact that laws are generally made by men to control women (Alexander, p. 268). He offers some general accounts as well: the primary function of the law is to preserve order (Alexander, p. 266), "to regulate and render finite the reproductive strivings of individuals" (Alexander, p. 267). What Alexander offers is, in short, a biological account of mores. He advances a causal account of moral dispositions and the patterns of moral behavior.

In commentary, Kenneth Schaffner argues that Alexander's theses (e.g., the hypothesis of the balance of power) may be too inexactly specified to allow them to be testable. Further, a multifactional account of human mores might prove more successful. In this regard, Schaffner defends a qualified emergentist account of social evolution. Though a complete evolutionary record, or knowledge of all the physico-chemical initial conditions and general laws of nature, might allow a reduction of social behavior to evolution, one must in the interim also make use of socio-cultural factors in explaining societal laws. Finally, Schaffner suggests that a successful socio-biology might have greater implications for normative ethics than Alexander admits. Such information would add to what we know concerning what humans and human societies are capable of, and concerning what humans are disposed to do. "A theory in normative ethics is rarely if ever fully insulated from factual implications" (Schaffner, p. 301). Socio-biological information concerning the human condition should have implications concerning what we take (and should take) to be the goods of human life.

This volume closes with a general commentary that seeks a different sense of foundation for the foundations of ethics and its relationship to the sciences. As Robert Solomon indicates, much of the discussion here concerning the relationship between science and ethics turned on the different roles of internal and external values (see, for example, the papers by MacIntyre, Toulmin, and Graham). In doing so, differing accounts were given of why one would engage in science or ethics. One has been offered, as Solomon indicates, a cultural justification (Dworkin), a historical justification (MacIntyre), a biological justification (Alexander), and a structuralist justification (Stent)[15] for ethics. Such approaches have, though, according to Solomon, failed to arrive at the core behind ethics, or behind ethics and science. "What links science and ethics together, whatever their differences, is their mutual demand for objectivity, what I have elsewhere called 'the transcendental pretense.' But science and ethics as objective practices have not yet been linked with science and ethics as personal engagements, whether motivated by the singular persuasiveness of a coherent culture or by some more individual psychological commitment which may or may not involve an explicit existential choice." (Solomon, p. 324). Solomon proposes a justification of these human practices by an account of what he terms the impersonal passions. He elaborates this suggestion with regard to science by forwarding the lineaments of the "emotional foundations of science" (Solomon, p. 329), drawing on what Polanyi termed the ambitious anthropo-centrism of our reason (Solomon, p. 329). The external and internal goods of science, the grounds for doing science, thus find their roots in "a passionate dedication to the search for truth in science. . . This passion *defines* science, through its internal rules and demands—impersonality ('objectivity') first of all, also *simplicity*, a voracious concern for 'the facts,' an uncompromising insistence on universal agreement and proof, but mainly, a demand for familiarity and understanding on a grand scale, using a relatively restricted scheme of explanation (for example, causal explanation). It is a mistake to think that all of this is simply part of 'scientific method'; if it were only that, we should have to repeat our question: why should anyone ever be motivated to take up that method? These demands are intrinsic to the passion itself' (Solomon, 327). In short, Solomon's commentary is an attempt at synthesis. He suggests how to go behind the traditional distinctions between the objective and subjective by looking for the roots of ethics and science in reasoned but impassioned motives. "The relation of science to ethics emerges clearly, not on the level of theories and codes themselves, nor merely at the level of their consequences or 'secondary links,' but right at the foundations, in the impersonal passions which constitute them" (Solomon, p. 329).

This volume and its essays thus offer a persistent analysis of the relations of ethics and the sciences. The approach is inter-disciplinary by choice and necessity: the interplay between mores and sciences, as well as among historical, conceptual, and scientific examinations of such interplays, transcends disciplinary boundaries. There are not only borrowings across disciplines, there are also examinations of issues that cannot be simply cast within the confines of a particular discipline. (e.g., Is Solomon doing philosophy or psychology? Is Alexander doing science or ethics?) Further, the interdisciplinary themes are, as suggested at the beginning of this introduction, complex. For example, in the course of this essay we have encountered the five interdisciplinary themes mentioned at the beginning of this introduction: 1) There is an interchange of ideas between science and ethics, which ideas have not only logical implications but metaphorical influences; 2) science can, at least in part, account for moral dispositions; 3) ethics has an empirical dimension, even if it is not grounded on facts (e.g., "evolutionary ethics"); 4) both science and ethics have concepts which are normative in character (i.e., there is no sharp line to be drawn between values and value-free facts); and 5) both scientific and ethical propositions are (in part, at least) contextual.

These five themes have served as leitmotifs in this and the previous volumes. Even if somewhat procrusteanly, they allow an overview of this series of discussions, and the place of this volume in it. Under the first rubric we have encountered circular journeys of ideas from evaluative to explanatory back to evaluative endeavors. One might think here of the journey of such concepts as 'organism' from social theory to biology, back to social theory, as indicated by Stephen Toulmin in his essay in the first volume, "Ethics and 'Social Functioning.' "[16] Biological concepts such as 'homeostasis' became transposed into social concepts by Durkheim and others leading, so Toulmin suggests, to viewing "Social nonconformity and political dissent as manifestations of social and psychological 'ill health.' "[17] In this volume, Graham offers glimpses of how approval of progress became approval of evolution, and how differing accounts of evolution then were taken to insure particular kinds of political programs. The point is that there is a conceptual interchange between the sciences and the understandings of social conditions, including evaluations of social conditions. Ideas feed back and forth between ethics and the sciences, taking on new significances, force, and functions.

Under the second rubric we have seen that a number of scientists have proposed explanations of ethical dispositions—scientific, descriptive ethics. In such arguments, an attempt is made to absorb ethics in whole or in part within the domain of science. Accounts are advanced as to how certain states of affairs are likely to dispose one to make particular moral evaluations. Ought is not derived from is, rather, an ought is caused by an is. Examples of this genre are Stent's suggestions in the second volume about certain ethical judgments being based on particular deep structures in human ethical reasoning and Richard Alexander's suggestions concerning the biological origins of social structures.

To this point in these volumes, we have had only one clear example of what could be termed the more ambitious variety of the third genre of argument concerning the relationship between ethics and science—Towers's paper, "Towards an Evolutionary Ethics."[18] There Towers argues that one can discern from evolution a social goal of a unified mankind, etc. Less ambitious arguments of this genre have simply contended that an adequate earthly ethics will need to take account of the lineaments of reality. Richard Alexander's paper in this volume probably belongs here.

The fourth point challenges the ability to develop, at least in some of the sciences, a value-free account of nature. The papers by Toulmin, Gorovitz and MacIntyre, and my own paper in the first volume support the fact that value-judgements are ingredient in the appraisal of physiological functions, the study of entities which must be regarded as particulars (e.g., porpoises, because such study requires judgments concerning the goods of porpoises), and the definitions of health and disease.[19] Issues of this sort are implicit in topics touched upon by Marcus and Margolis—talk about mental functions as normal or abnormal requires more than value-neutral descriptions of behavior. Or, rather, any full description of human behavior is already value-infected.

The last point turns on the meaning of both science and ethics—whether reason-giving is of one fabric, or whether it breaks into many islands of dissimilar discourse where what counts as a successful appeal to reason or failure of such an appeal is decided by different criteria. This question arises both within science and ethics, as well as between them. Somewhat as Ludwik Fleck, Thomas Kuhn, and others have argued, the sense of reason-giving may to a great extent be isolated within the domains of particular thought collectives or paradigms. Hauerwas and Burrell extended this argument to ethics in last year's volume.[20] In this volume, Dworkin's critique of the storyless autonomous individualist resumes the Hauerwas-Burrell argument in part by questioning the scope of the validity of claims made in ethics. Vlastos's Plato may belong here as well, since for Vlastos's Plato, the individual moral agent has a place only within the context or ambience of the polis. Finally, the Cassell-Ramsey essays concern the coherence of the moral agent. Cassell raises the question of whether the self itself is contextual, of whether we do not ourselves have more than one story.

This fifth theme involves the context of scientific accounts as well: the psychological and social factors that fashion a particular scientific community and its world view. Toulmin's paper explores these issues by examining the motivation of scientists. The same factors appear in Graham's paper on a more sociological scale. In order to demonstrate what will make one or another scientific view more persuasive, one must, as Graham shows, attend to historical accidents, to how stories get told, to how scientific views fit into general views of reality. Prevailing value assumptions help shape the accepted account of reality. The other side of this point surfaces when one asks, what are the consequences of holding certain theories to be correct—an issue raised by Marc Lappé in the first volume of these papers.[21] This is also raised by Graham when discussing the appraisal by persons in the 1920s of the consequences of adapting Lamarckian or Mendelian genetics. In short, scientists act out of a moral background or ambience, and science in turn affects that ambience. The circle is complete. Moreover, as Alasdair MacIntyre argues in this volume, the circle may be completed in unexpected ways. Science may in the end be best understood as a morality. Science may be comprehended only by reference to certain internal goods (e.g., commitment to realism) that define the modern practice of science. The context of science would then be found within itself— in its own ingredient goals.

The articles in these three volumes have thus sketched how ideas can be borrowed from social theory by science and then be given back to social theory, and how the foundations of value-theory and of science in this fashion exchange concepts. It has been shown that science both presupposes and influences evaluations. Moreover, sciences in various degrees and ways have an evaluational component, and ethical and scientific reasoning is always to some extent context bound. The interdependencies between the foundations of ethics and the sciences are thus complex. Because of their complexity it is useful to characterize these interdependencies as conceptual, causal, and contextual.

(a) Conceptual bonds exist between science and ethics as demonstrated by the first theme concerning the borrowing of concepts, the third theme concerning the possibility of reading value-judgments from states of nature, and the fourth theme concerning the ingredience of value-judgments in the sciences.

(b) There are causal bonds as the second theme indicates: there may be biological bases for ethical systems.

(c) Actual ethical and scientific judgments are dependent upon context. The sixth theme concerning the psychological context of science and the contextual nature of ethical reasoning belongs here. With some appropriate transformations, one may also be

able to place here the third theme concerning reading ethical laws
from the tendencies of nature; that is, since concrete ethical
judgments are dependent upon context, they require a real world
with a particular content.

These clusters of issues indicate ways in which ethics and the
sciences can be mutually illuminating. They show areas where
the methods of science can help account for evaluations, and
where ethics and the sciences share basis subject areas in common. They spring from an interest in a coherent view of ourselves and our circumstances. Here we find the "impersonal
passions" of the interdisciplinarian to have similarities with the
"impersonal passions" of the librarians in Borges's library. We
are driven to comprehend, as best we can, the immense compass
of reality in human terms. Though this task is open-ended, often
fraught with misguided presuppositions, and frequently evocative
of despair, it is the most noble of human enterprises. It is the
attempt to define or discover our condition as a human condition,
a reality not alien to our goals and needs for coherence.

NOTES

1. Jorge Luis Borges, *Ficciones* (Buenos Aires: Emece Editores,
   1956), pp. 75, 85.
2. For this point one need not take as idealistic a view as did Kant,
   that the order of the phenomenal world is our construction.
   However, it is the case that in knowing the world we impose our
   expectations upon it. Norwood Hanson, *Patterns of Discovery*
   (London: Cambridge University Press, 1965).
3. This point is Kantian, namely that both the world of science and
   that of moral conduct have the character of being bound by universal laws which must be transparent to thought, if the world is to
   have coherence for us. Compare, for example, Kant's *Fundamental
   Principles of the Metaphysics of Morals* (1785) with his *Metaphysical Foundations of Natural Science* (1786). Or, consider Kant's
   regulative employment of reason: "In order, therefore, to secure an
   empirical criterion we have no option save to presuppose the systematic unity of nature as objectively valid and necessary." Immanuel Kant, *Critique of Pure Reason*, trans. Norman Kemp Smith
   (London: Macmillan, 1968), p. 538: A652-B680. In short, as rational knowers and doers we must construe our knowing and doing

in rational terms. As a consequence, one would expect foundations common to both ethics and science.

4. One might think here of the Hegelian contention that in true knowing there is a full coincidence of thought and being. Being must be adequate to thought, as thought is to being. Thus where one recognizes the structures of thought in nature, it is as true to say that they are there, as that we imposed them. As Hegel put it, the point of the philosophy of nature is to find "only the mirror of ourselves, to see in nature a free reflection of spirit: to understand God, not in the contemplation of spirit, but in this His immediate existence." G.W.F. Hegel, *Philosophy of Nature*, trans. M.J. Petry, vol. 3 (London: George Allen and Unwin, 1970), p. 213, Zusatz S 376.

5. Joseph Margolis, "The Concept of Disease," *The Journal of Medicine and Philosophy*, 1 (September 1976): 238-55; and H. Tristram Engelhardt, Jr., "Ideology and Etiology," *The Journal of Medicine and Philosophy*, 1 (September 1976): 256-68. H. Tristram Engelhardt, Jr., "Human Well-Being and Medicine: Some Basic Value-Judgments in the Biomedical Sciences," in *Science, Ethics and Medicine*, ed. H.T. Engelhardt, Jr. and Daniel Callahan (Hastings-on-Hudson, N.Y.: Hastings Center, 1976), pp. 120-39.

6. "Are the principles of right living really connected with the intimate constitution of the universe? . . . Now science began to return a decisively affirmative answer to such questions as these, when it began, with Mr. Spencer to explain moral beliefs and moral sentiments as products of evolution. For clearly, when you say of a moral belief or a moral sentiment that it is a product of evolution, you imply that it is something which the universe through untold ages has been laboring to bring forth, and you ascribe to it a value proportionate to the enormous effort that it cost to produce it." John Fiske, *Evolution and Religion in Excursions of an Evolutionist*, in *The Miscellaneous Writings of John Fiske*, vol. 7 (Boston: Houghton, Mifflin and Company, 1902), pp. 276-77.

7. Aristotle, *Politics* 1.5.1255a.

8. Ludwik Fleck, *Entstehung und Entwicklung einer wissenschaftlichen Tatsache* (Basel: Benno Schwabe, 1935).

9. See, for example, Thomas Kuhn's acknowledgement of his indebtedness to the work of Ludwik Fleck, *The Structure of Scientific Revolutions* (Chicago, Ill.: University of Chicago Press, 1962).

10. Fleck, *Entstehung und Entwicklung*, pp. 107, 108.

11. David Burrell and Stanley Hauerwas, "From System to Story: An Alternative Pattern for Rationality in Ethics," in *Knowledge, Value*

and Belief, ed. H. Tristram Engelhardt, Jr. and Daniel Callahan (Hastings-on-Hudson, N.Y.: Hastings Center, 1977).

12. Michael Polanyi, *Personal Knowledge: Towards a Post-Critical Philosophy* (Chicago, Ill.: University of Chicago Press, 1958).

13. Paul Feyerabend, *Against Method* (Atlantic Highland, N.J.: Humanities Press, 1975).

14. Vlastos signals Plato's defense of the rights of women and of the equal consideration of women in the *Republic*. As he shows, this is in marked contrast with the less extensive rights of women, though still more extensive than in Greece of that day, as presented in the *Laws*, which relied upon the oligarchic principle of proportional equality.

15. Gunther Stent, "The Poverty of Scientism and the Promise of Structuralist Ethics," in *Knowledge, Value and Belief*, ed. H. Tristram Engelhardt, Jr. and Daniel Callahan (Hastings-on-Hudson, N.Y.: Hastings Center, 1977), pp. 225-46.

16. Stephen Toulmin, "Ethics and Social Functioning: The Organic Theory Reconsidered," in *Science, Ethics and Medicine*, ed. H. Tristram Engelhardt, Jr. and Daniel Callahan (Hastings-on-Hudson, N.Y.: Hastings Center, 1976), pp. 195-217.

17. Ibid., p. 217.

18. Bernard Tower, "Toward an Evolutionary Ethic," in *Knowledge, Value and Belief*, ed. H. Tristram Engelhardt, Jr. and Daniel Callahan (Hastings-on-Hudson, N.Y.: Hastings Center, 1977). pp. 207-24.

19. Stephen Toulmin, "Ethics and Social Functioning," pp. 195-217, Alasdair MacIntyre and Samuel Gorovitz, "Toward a Theory of Medical Fallibility," pp. 248-74, and H. Tristram Engelhardt, Jr., "Human Well-being and Medicine: Some Basic Value-Judgments in the Biomedical Sciences," pp. 120-39. All in *Science, Ethics and Medicine*, ed. H. Tristram Engelhardt, Jr. and Daniel Callahan (Hastings-on-Hudson, N.Y.: Hastings Center, 1976).

20. David Burrell and Stanley Hauerwas, "From System to Story: An Alternative Pattern for Rationality in Ethics," in *Knowledge, Value and Belief*, ed. H. Tristram Engelhardt, Jr. and Daniel Callahan (Hastings-on-Hudson, N.Y.: Hastings Center. 1977), pp. 111-52.

21. Marc Lappé, "The Non-neutrality of Hypothesis Formulation," in *Science, Ethics and Medicine*, ed. H. Tristram Engelhardt, Jr. and Daniel Callahan (Hastings-on-Hudson, N.Y.: Hastings Center, 1976), pp. 96-113.

1

Objectivity in Morality and Objectivity in Science

Alasdair MacIntyre

ETHICS AND POLITICS are outlying territories on the presently conventional map of the discipline of philosophy, provinces inhabited and worked, so the fashionable say, by worthy enough persons, but not to be compared for importance or glamour with the metropolitan centers of logic, semantics, epistemology, and philosophy of science. Indeed, the suggestion is sometimes even made—in the coda to Dummet's book on Frege, for example— that it is only when the major problems have been solved in metropolis that the techniques will become available which will enable the peasants in the outlying districts to solve their problems too. Those of us who live and work in these outlying territories have a number of reasons for exhibiting provincial recalcitrance to these metropolitan claims. For one thing, the claim that a great new technical advance is about to solve the problems of the metropolitan areas has by now been made just a little too often. But more importantly, it has become clear recently that to some degree the philosophy of science is now recapitulating the history of ethics and politics (for convenience' sake, I shall from now on abbreviate "ethics and politics" to "ethics") and has even become a little self-conscious about so doing, perhaps because of a certain uneasiness about the suggestion conveyed by Santayana's epigram that those who recapitulate the history of philosophy do so because of their own ignorance.

I want to undertake three tasks in this present essay. First, I want to set out the parallels between ethics and the philosophy of science a little more precisely than has been done elsewhere so far. Second, I want to identify, if possible, the causes of this parallelism or recapitulation. And third, I want to suggest that the philosophy of science may have something to learn from ethics, rather than vice versa.

The parallels between philosophy of science and ethics are historically rooted in the parallel structures of Kant's first and second *Critiques*. For what each of the *Critiques* did was to detach a set of universal generalizations from a background context which had hitherto provided their justification and their intelligibility. The key figure in the seventeenth-century background to the *Critiques* is, from one point of view, Robert Boyle. For it was Boyle who successfully transplanted the notion of *lex naturalis* from ethics to science without abandoning the theological background of that notion. God has purposes for both man and nature; his instrument for achieving these is a law which he utters both to men and to planets. The difference between men and planets is that planets are never disobedient. The task of the natural scientist, like that of the moralist, is to decipher and reconstruct the *lex naturalis*. From this point of view, Newton's achievement in the *Principia* has an entirely fitting sequel in his later interpretations of the *Book of Daniel*; they are simply two stages in the same inquiry. Kant's great if unintended achievement was to make the intellectual unity of Newton's life unintelligible to later generations.

In Renaissance science, of which Newton's work was the culmination, it had of course been metaphysical theology that linked each part of the universe to the whole and that therefore allowed experiment to vindicate generalization. When the metaphysical premises are rejected—and empiricism had preceded Kant in rejecting them—then we are left with a stock of universal generalizations on the one hand, a set of particular experiences on the other, and the puzzle of how to relate the one to the other. Empiricism accepted the Quixotic task of solving this impossible puzzle, and two hundred years later empiricists still hope that modern improvements in the sophistication of their logical weaponry will enable them to defeat the windmills at last. But Kant showed, and it should have been once and for all, that experience can provide no foundations for the generalizations of physics any more than for those of ethics. Yet, it was the effect of his work to accentuate the differences and to obscure the kinship between physics and ethics. By the particular way in which Kant made both physics and ethics independent of metaphysical theology, he prepared the way for a *Weltanschauung* in which all connection between physics and ethics disappeared from view. "Science" was one thing, "morality" quite another. The most prosaic categories of the distinctively modern mind thus embody a vulgarized and unacknowledged version of Kant. Nonetheless, both physics and ethics continued to be marked by their shared, if forgotten, inheritance. For each has continually found itself confronted by the same task: how to find a status for its own set of universal generalizations which will not violate the now internalized Kantian prohibitions.

It is true that modern ethics was forced to accept the detachment of its universal generalization from any attempt to force them upon experience long before modern philosophy of science did, although a hundred years of utilitarian resistance witness to how difficult it has been to learn what Kant has to teach. But already in the nineteenth century, moral theorists as different as Kierkegaard and Emerson were responding to Kant in a radical way that their contemporaries in the philosophy of science never paralleled. It is perhaps for this reason that ethics has shown a striking tendency to anticipate the philosophy of science in its commission of error, so that contemporary philosophers of science often remind us of nineteenth-century moralists. Consider, for example, Kuhn's reincarnation of Kierkegaard, and Feyerabend's revival of Emerson—not to mention the counterpart to them all, in which the eighteenth century outbids the nineteenth, Polanyi's version of Burke.

Let me begin with *Either/Or* (Copenhagen: 1841) and *The Structure of Scientific Revolutions* (Chicago: 1962). Kierkegaard had learned directly from his reading of Kant that the universal generalizations which formulate the categorical imperatives of the ethical ("Always tell the truth," "Always keep your promises," etc.) could not be derived as conclusions either from premises about experience or from premises in metaphysical theology. He taught himself—or perhaps learned from Hegel—that Kant's rational test for the categorical imperative failed. He was well aware that there are rival sets of universal generalizations prescribing ways of life to us. If we cannot decide between them by appeal to experience or indeed by any other mode of rational argument, then it seems to follow that the only way to decide between them is by a fundamental, unargued, and unarguable choice in the making of which there are no rational criteria to which appeal can be made.

Kuhn, too, is preoccupied with the question of how decisions are to be made between rival sets of universal generalizations. It is true that Kuhn locates the need for radical choice at some points in the history of science rather than others; but the choice that is implied by his account is radical in just the way that Kierkegaard's is. It is a choice of first principles, and therefore cannot be further argued for; for it itself determines which arguments have weight and which do not. It follows that the natural scientist who lives in a period of transition between paradigms must at some point in his career make precisely the type of arbitrary, criterionless either/or choice which Kierkegaard describes. Moreover, he follows Kierkegaard beyond *Either/Or* into the *Concluding Unscientific Postscript* in his relativism concerning ontology: "There is, I think, no theory-independent way to reconstruct phrases like "really there"; the notion of a match between the ontology of a theory and its 'real' counterpart in nature now seem to me illusive in principle" (p. 206 of Kuhn's concluding unscientific *Postscript* to the 1969 edition of *The Structure of Scientific Revolutions*). Kuhn does not want to call his position relativistic, but then Kierkegaard would clearly have disowned this description too. Nonetheless, Kuhn's resistance to this type of interpretation of his position ought in fairness to be noted. Kuhn has insisted vehemently that Lakatos's reading of him as an irrationalist is a gross misreading, and he could make the same case for misrepresentation against me that he made against Lakatos, unless I allow that he does indeed insist that the progress from paradigm to paradigm must be rational. Only, like

Lakatos, I cannot see that this claim is in any way consistent with important passages in Kuhn's text.

Consider now a second type of example. Kant's detachment of universal generalizations from their metaphysical context on the one hand and their foundation on experience on the other raises the question: Who is to decide which are the authentic universal generalizations? For Kant himself the answer is easy. The rational agent prescribes the maxims which express the universal generalizations to himself, whether it be in arithmetic, in physics, or in morality. And since reason is the same in all individuals, all individuals will legislate the same set of maxims. This autonomy of the rational individual is to be contrasted with the false heteronomy of the traditional churches. It is this autonomy which Emerson learns from Kant, but he rejects at the same time any conception of a universal reason. Every individual has to decide for himself, to think for himself, to feel for himself: "the Distinction of the new age" is "the refusal of authority." (*Journals*, 4:495.) Every individual must judge for himself in what progress consists. Santayana wrote of the Emersonian transcendental philosophy that it "enables a man to renovate all his beliefs, scientific and religious, from the inside, giving them a new status and interpretation as phases of his own experience or imagination; so that he does not seem to himself to reject anything, and yet is bound to nothing except his creative self." (*Character and Opinion in the United States*, New York: 1920, p. 13). Emerson's rebellion against authority was, of course, not a rebellion against an authentic entrenched tradition, such as that of the Roman Catholic hierarchy. The authority he attacks is the liberalized Unitarianism of his own father, which was, in turn, a rebellion against the already liberal Unitarianism of the preceding generation; and that preceding generation had itself overthrown what they looked back to as the tyranny of congregationalism. So in our own time the Emersonians of the contemporary philosophy of science rebel against the liberalized Popperianism of Lakatos which was, in turn, a rebellion against the already liberal Popperianism of Popper and his first followers; and they, of course, had overthrown what was looked back to as the tyranny of the Vienna Circle: "let us free society from the strangling hold of an ideologically petrified science just as our ancestors freed *us* from the strangling hold of the One True Religion!" (*Against Method*, London: 1975, p. 307). "Incidentally it should be pointed out that my frequent use of such words as 'progress', 'advance', 'improvement' etc., does not mean that I claim to possess special knowledge about what is good and what is bad in the sciences. . . . Everyone can read the terms in his own way. . . ." (Ibid., p. 27). We need "a form of thought that 'dissolves into nothing the detailed categories of the understanding', formal logic included." (Ibid., p. 27, quoting Hegel, *Wissenschaft der Logik*, Hamburg: 1965, I:6).

So writes Feyerabend, reproducing Emersonian individualism, concealing his moralistic ancestry by citing anarchism and Dada, but revealing it in passage after passage. Yet if Kuhn has resurrected Kierkegaard and Feyerabend reincarnated Emerson, both are in fact reacting to—or at least write as if they are reacting to (for they never acknowledge any stimulus, let alone any debt)—an earlier philosophy of science which sets the scene for them. I mean, of course, Michael Polanyi's. There are two key elements in Polanyi's thought to which Kuhn or Feyerabend or both have unacknowledged debts. The first is his stress upon the way in which the appeal to experiment and observation operates only under normative constraints. The result is that theory is protected from certain kinds of falsification and that the stability of large-scale theory resembles the stability of systems which modern Western thought takes to be nonrational, such as Zande witchcraft. Hence the competition of theory with theory takes place for the most part within a protected area. This account of the continuity and stability of science yields as its obverse Kuhn's account of science's discontinuities and instabilities. It yields also the same kind of conclusion about the relationship of theory to observation which Kuhn formulates in terms of incommensurability.

A second key element in Polanyi's account is his view of the roles of consensus, tradition, and authority in the scientific community. When Feyerabend objects to institutionalized natural science as an oppressive ideological force, he seems to presuppose a view almost identical to Polanyi's; the tone is, of course, very different, but different in just the way that an Emersonian account of England in the eighteenth century would differ from Burke's. Feyerabend's philosophy of science is indeed Polanyi's turned upside down, while Kuhn's is a simple—and vulgarized—adaptation.

This pedigree makes the parallel with nineteenth-century religious moralities intelligible. For Polanyi's achievement was to exhibit the community of natural scientists as a community of faith, a church carrying within it its own developing tradition. This type of semiliberal, often Anglican, theology, to which his philosophy is a counterpart naturally breeds individualist reactions of an existentialist or an Emersonian kind. But Polanyi, of course—like Burke—combined with his emphasis on consensus and tradition a deep commitment to a realistic interpretation of science. Polanyi's realism rested on what he called a "fiduciary commitment." Feyerabend (and less explicitly Kuhn) have retained the fideism; what they have rejected is the realism and with it the objectivism which Polanyi held to as steadfastly as any positivist.

What went wrong? How did subjectivism and relativism achieve these victories first in ethics and the philosophy of religion and then in the philosophy of science? Let us return to the Kantian starting point and consider how easily its objectivism can be imperiled. Withdraw the Kantian account of universal reason from either physics or ethics or religion and what is left is an isolated, autonomous, socially contextless individual confronting the rule-governed practices of the physicist, the moral agent, and the religious believer. Since the rules can no longer be vindicated by objective, impersonal reason, and since experience no longer seems to provide a neutral court of appeals, two and only two relationships between the individual and the rules seem to be possible. Either I can accept or reject the rules by my own free, unconditioned, criterionless choice; or the rules can be imposed upon me by some external power or influence.

This crude and simple picture of the relationship of the individual to rules has been extraordinarily influential. Elsewhere, I have traced its history in social theory from Durkheim and Weber through to Dahrendorf and Goffman. (In *After Virtue*, forthcoming). Here, I want to turn immediately to a diagnosis of the source of the error that it embodies. That error lies in a misunderstanding of the relationship between what I shall call rules of practice and what I shall call institutional rules.

By a rule of practice I shall mean a rule whose point and purpose is derived from its role within a form of activity which has goods internal to it. Examples of such forms of activity or practices are chess, natural science, and painting. Internal goods are defined by contrast with goods external to a practice. Consider, for example, the case of a small but intelligent child to whom we wish to teach chess. The child has no interest in learning to play chess, but does have an insatiable appetite for candy. We therefore promise the child fifty cents worth of candy each week, if the child will play chess with us. Moreover, we tell the child that we shall always make it difficult, but not impossible, for the child to win, and that if the child wins he or she will receive an additional ration of candy. So long as the child is playing chess only for the sake of the candy, the child has, of course, no reason not to cheat, if he or she can cheat successfully. For playing chess is at this stage only a means to the acquisition of an external good, the candy. But if and when the child comes to appreciate the goods internal to chess, then the rules acquire a new kind of authority. For not to play according to the rules is to deprive oneself of a share in those goods which give the rules their point; it is to defeat oneself and not one's opponent.

External goods, then, are those contingently related rewards—candy, money, reputation, status, power—which may derive from successful participation in a practice; internal goods are those achievements of excellence which exhibit human aesthetic, imaginative, intellectual, and physical powers at their highest. External goods are always the property of some particular individual and the conditions of human life are such that characteristically the more one person has of such goods, the less there is for others. Internal goods are always both a product and a possession of forms of community, even when it is a particular individual, a Turner in painting, or a Planck in physics, who excels. There is no way to recognize internal goods except by participation in practices, even although many theorists deny (Freud in his theory of art) or ignore (Hume on the virtues, Rawls on primary goods) their existence altogether. Nonetheless, it is they alone which confer authority on the rules defining the practice, the rules without which the goods internal to the practice cannot be achieved.

When the concept of an internal good is lost sight of, then it becomes impossible to distinguish the rules of a practice from the rules of those institutions which are the bearers of practices. Chess, scientific inquiry, and painting are examples of practices; chess clubs and federations, universities and laboratories, galleries, museums, and art dealers are examples of corresponding institutions. Institutional rules may be justified in a number of ways, but they possess authority insofar as they embody or support rules of practice; and to assimilate practices to institutions leads all too easily to a picture of the evaluative world in which the individual conceived as autonomous and socially contextless (except accidentally) is judge and arbiter, while all social relationships are conceived as merely "positive." No positive set of rules has any authority, except and insofar as the autonomous individual chooses to confer it upon them. The individual's moral identity is prior to, and independent of, any of his or her social relationships; all relationships are a matter for individual criticism and judgment.

In the context of a practice, however, the individual is not generally or usually a judge or an arbiter; he or she is a participant in acknowledging an authority whose character has emerged in the history of the practice in question. Hence, the individual's moral identity derives from his or her having found a place, no matter how insignificant, in that history; the individual has to recognize him or herself as having a subordinate part, no matter how eminent, in the more than individual projects that constitute the practice. It is this subordination of individual experience and thought that supplies the crucial element of impersonality and objectivity to practices.

Examples of practices in which such authority and such objectivity have been crucial are law, painting, and natural science. Within each of these practices there have been sustained tasks and projects, which have gradually emerged over long stretches of time. The elaboration of conceptions of liability, responsibility, and negligence has been one such project, the painting of the human face another. It would of course be easy enough to write the history of law or the history of painting, so that the distinctions of the conceptual schemes embodied in codes (Roman law as against Anglo-Saxon) or the distinction of styles (Fra Angelico as against John Singleton Copley) led to a formulation of incommensurability theses of a kind which would make the continuities of such projects incomprehensible—and Feyerabend, of course, has seen in such a possibility support for his own position in the philosophy of science (*Against Method*, chap. 17). But this would not only disrupt our writing of history; it would conceal the purposes of art and law which transcend styles and codes. As in art and law, so also in natural science. And, like art and law, natural science is a set of projects which embodies a moral task. That moral task is partially but importantly defined by the commitment of science to realism. Kuhn has written that "A scientific theory is usually felt to be better than its predecessors not only in the sense that it is a better instrument for formulating and solving puzzles, but also because it is a better representative of what nature is really like." Kuhn's acceptance of incommensurability leads him to repudiate realism. What this repudiation brings into view is the connection between realism on the one hand and a certain view of the continuity of science on the other. Where we stand on the issue of realism and how we write the history of science are questions which have to be answered together or not at all. This connection has been obscured by an inadequate understanding of how realism ought to be construed.

Realism was originally an epistemological thesis about what the scientist (or indeed anyone at all) knows when he knows that such and such is the case. It becomes in its more recent incarnations—and indeed in any version must be—a semantic thesis about the sense and reference of the key terms in scientific theories. Hilary Putnam, who treats it as a semantic thesis, also proposes to treat it as any hypothesis; we conjecture that the most probable explanation of the fact that science represents nature as composed of protons and μ-mesons and the like is that nature really is composed of protons, μ-mesons, and the like.

The incommensurability thesis is clearly at odds with realism thus (or perhaps anyhow) construed, as Kuhn has explicitly recognized. The question is: at what level ought the disagreement between them to be resolved? Over what kind of issue is the conflict most crucial? Kuhn and Feyerabend clearly treat the controversy as *primarily* semantic. They try to demonstrate incommensurability by appealing to differences in meaning; and they argue for certain consequences of incommensurability from premises concerning just such differences. This view of the primacy of issues of meaning is itself unargued by either author and it is thoroughly in accordance with the *Zeitgeist*, although *Geist* has perhaps moved just a little bit recently beyond that particular *Zeit*.

I want to suggest a quite different perspective; one that would put questions of meaning in a not unimportant but subordinate place. If I were to assert—correctly—that the word "acid" has changed its meaning in the last two hundred and fifty years and someone were to inquire in what way, the right reply would be to recount parts of the history of chemistry. Changes in meaning are part of changes in theory; continuities and discontinuities in meaning are part of the continuities and discontinuities of theories and other schemes of belief or conjecture. Any theory of change of meaning that violates our historical understanding of the continuities—or the discontinuities—of theories thereby give us grounds for rejecting it. What in such a perspective are we to make of realism?

It will be conceived only secondarily as a semantic thesis and not at all as an hypothesis. Rather, it will be regarded as what Kant called a regulative ideal. What did and does it regulate? The scientists' interpretation of their own task. It is an ideal which is at home in other spheres, in theology, for example. Bellarmine, who urged instrumentalism as an interpretation of science upon Galileo and others, was a thorough-going realist in theology. But since Galileo, realism has been the ideal which at once sets constraints upon what is to count as the solution of a scientific problem and provides an interpretation of scientific results. Of course any particular piece of theory, detached from the context of scientific activity, can be interpreted in a number of ways, realist and non or antirealist. But the practice of science through time presupposes a continuous adherence to realist goals, whatever side-turnings there may be on the way to acheving them.

This will be obscured if we abstract from the history of science as a whole parts of fundamental physical theory and try to write their history in isolation. The leap from classical mechanics to either quantum mechanics or special relativity is easily portrayed as having the same kind of discontinuity that seems to have been present in the leap from Heraclitus to Parmenides. But the history of fundamental physical theory is always only part of a longer story. As Lenin rightly saw, one key way to write the history of science is to make cosmology the central scientific discipline, and cosmology is realistic in just the way that history itself is realistic. In writing cosmology, after all, we are writing the prologue to human history and thus rewriting (in part at least) Genesis I. Philip Gosse, who, like Bellarmine, was a realist in theology, would have been delighted with Kuhn's attack on any connection between natural science and realistic ontological claims.

But obviously it is not only because every part of physics and chemistry has what status it has as a potential or actual contributor to the realistic narrative of the cosmologist that there are continuities in science which escape us if we give a false place to fundamental physical theory. For it is above all in middle-level theory that continuities are maintained, just as it is at this level that the important convergences betweeen different branches of science appear. In middle-level theory the characteristic structures are those of such enterprises as the kinetic theory of gases or the theory of the chemical elements. Fundamental theory helps to provide changing answers to the same relatively unchanging questions of middle-level theory; and the continuity of the questions puts constraints on the degree of discontinuity that we can ascribe to the answers.

Consider three types of sequence in the history of nineteenth- and early twentieth-century science. The first is in the scientific career of Henry G.J. Moseley. In 1913, when Moseley designed his experiments on X-ray spectra, he chose a series of twelve chemical elements for his investigation because they were a continuous series in the table of atomic weights as formulated by Mendeleev; his experiments led him to the conclusion that the table had to be revised—the places of cobalt and nickel were reversed—because the regularity of the table derived not from atomic weights but from "a fundamental quantity, which . . . can only be the charge on the central positive nucleus." Thus, within one single continuous investigation, recorded in one paper *The High Frequency Spectra of the Elements, Part I (Philosophical Magazine,* December 1913), Moseley uses the vocabulary both of Mendeleev and of Bohr, the language of 1863 and the language of 1913, without any obvious signs of linguistic strain or change of meaning.

A second longer sequence, in which the first would have to appear as an episode, is the history of the kinetic theory of gases. The kinetic theory has undergone many transformations in its long history from the first anticipations by Bernoulli, through its founders, the unfortunate Waterston, Joule, Rankin and Kronig, and Clausius, to Maxwell, to Boltzmann, and finally to its modern quantum mechanically based expositors. But it is recognizably *one and the same theory,* and its continuities are brought out in the way in which earlier insights are put to new uses or help to set new problems. So a straightforward modern textbook exposition of the theory, like Walter Kauzmann's, is historically layered; the contemporary theory resembles not a twentieth-century building on the site of earlier buildings which it has replaced, so much as an originally eighteenth-century mansion, rebuilt in parts, with the foundation replaced, with many additions and much ornamentation, but still recognizably the same building for which, therefore, the original architects must be given credit even before those responsible for the most brilliant parts of the reconstruction.

Wherein does the unity of the history lie? In the continuous attempt to construct a realistic representation of the inner structure of gases which will enable us to diagram the relationship of the microproperties of gases to the nature of the molecules of which gases are composed. The realism is part of the content of the theory: the use of Van der Waal's equation has always been explanatory and not just calculative, making the kinetic theory not (as has sometimes been suggested) a theory of ideal gases, but rather an ideal theory of real gases. The introduction of quantum mechanical theory achieved at least two things: it *confirmed* the general character of the hypotheses relating the character of gases to that of molecules and the character of molecules to that of atoms, and it did so in part by eliminating from the theory the paradoxical consequences which Boltzmann had derived from the late nineteenth-century versions of the theory. In 1890 Boltzmann had shown that on the basis of the assumptions of classical mechanics plus the assumption that atoms have an internal structure, have parts, the specific heat of any ordinary piece of matter should be far larger than it is; the actual specific heat can by contrast be accounted for solely by the motions of the atoms composing it. Boltzmann thereby helped to generate a scepticism about the kinetic theory which he himself never shared, a scepticism that was only finally defeated by Bohr's theory of the internal structure of the atom. We ought to note that Boltzmann's defense of the kinetic theory against the sceptics was vindicated by Bohr's theory, and by Einstein's predictions of 1905, only because both Einstein and Bohr presented a realist account of the internal structure of the atom. Planck's instrumentalist account of 1900 could not have solved Boltzmann's problem.

In this sequence of events that runs from the 1840s to the 1920s we can see writ large over eighty years the continuities that already appeared within one paper by Moseley. The concepts of a gas, an atom, an element, a molecule, internal structure, ray deflection, atomic weight, electrical charge persist. The table of the elements is corrected, completed, and extended. Physicists and chemists who lived through the period saw the questions of their youth—questions that come to a head in papers like Maxwell's on Boltzmann or Lord Kelvin's 1901 paper on *Nineteenth Century Clouds over the Dynamical Theory of Heat and Light*— replaced by the answers of their old age—unless, like Boltzmann, they had committed suicide (in part because of his depression over scepticism about the kinetic theory) or, like Moseley, had been killed in the First World War.

Consider now a third related sequence, that in which at the most fundamental level physicists replaced classical mechanics by quantum mechanics and by special relativity. It is certainly here that we find the most radical forms of conceptual change and innovation; *mass* is, after all, redefined and what could be a more fundamental change than that? If, then, we strip away from fundamental physics thus described that whole scientific context in which it has its setting, we shall indeed discover *incommensurability* in the pure form described by Kuhn and Feyerabend.

Of course to maintain incommensurability we shall have to re-describe even the history of fundamental physics in some surprising ways. It will turn out, for example, that where physicists thought that they had performed only one set of experiments they had in fact performed two. This is not just my view; it is a consequence which Feyerabend himself draws (*Against Method,* pp. 282-3) for the Michelson-Morley experiment, as well as for the experiments concerning the variations of the mass of elementary particles or the transverse Döppler effect. Feyerabend stresses quite correctly the fact that these experiments have to be given a relativistic interpretation in order to provide support for special relativity, and that, in order to understand why they undermine classical mechanics conjoined with electrodynamics, we have to describe them in classical terms. But from this he infers that the experiments, differently interpreted and described, constitute two sets of experiments and not one. "Why should it be necessary to possess terminology that allows us to say that it is the same experiment which confirms one theory and refutes the other?" (Ibid., p. 282). To the objection that his own account presupposes the very identity which he denies, Feyerabend replies that *his* language is one in which the expressions of relativistic and of classical mechanics are not used, but referred to: "The language in which *this* discourse is carried out can be classical, or relativistic, or Voodoo." (Ibid., p. 284). The force of this reply depends upon the notion that relations of meaning in natural languages must be elucidated in terms of a theory of types applicable to formal calculi, or, if it does not, I can make nothing of Feyerabend's use and mention of the use/mention distinction. But there seems no reason to accept this assumption; Feyerabend's language in the only relevant sense is English and English is *neither* classical, *nor* relativistic, *nor* Voodoo, but allows one to speak of and to relate to each other theories using very different terms. Indeed the notion of "the language" in Feyerabend is extremely unclear.

What is clear, however, is that some considerations drawn from a largely unstated theory of meaning are being used to support conclusions about the *identity* of experiments and the *continuity* of theories. Yet it is also clear that the history to which these considerations are being applied is a diminished and a partial history. It is a history in which the most abstract theoretical concepts of physics have been placed in the foreground, with the structures of theoretical chemistry on the one hand, and the concrete experiments of both physics and chemistry on the other, put in secondary positions. Questions of meaning are to be settled with regard to the abstract physical concepts considered in isolation in the first instance, and then whatever consequences need to be drawn for chemistry or for the interpretation of experiment are to follow afterward. But there seems no more reason to give this kind of primacy to abstract theoretical concepts in respect of meaning than there was to agree with the positivists in giving primacy to observation statements.

Moreover, the weight accorded to questions of meaning itself seems unwarranted. We possess at present no good overall theory of meaning. From Frege till now the attempts to construct general theories have produced many philosophically important concepts and theses; but no general theory has been able to sustain its claims, partly because the elementary problem of adequately representing natural language in some formalism (as contrasted with constructing illuminating ad hoc formalizations of this or that part of some natural language) has never been solved. Davidson's gallant and exciting program, for all the work done on it, still remains just that: a program.

Feyerabend is of course in the right when he sternly reproves those who seek to avoid questions of meaning altogether; but what will not do is to make the well-established continuities of history unintelligible by invoking much less well-established theories about untranslatability (Ibid., chap. 12, passim). (We ought to note in passing that Kuhn and Feyerabend can draw no support from such theses as have been cogently argued in this area, most notably Quine's, since Quine's thesis extends to every matching of sentence with sentence and thus would not allow some matchings rather than others to be untranslatable, as the *incommensurability* thesis requires).

I therefore take it that we have been given no good reason for resisting the contentions that history has primacy over semantics and that the continuities of history are moral continuities, continuities of tasks and projects which cannot be defined except with reference to the internal goods which specify the goals of such tasks and projects. Those tasks and projects are embodied in practices, and practices are in turn embodied in institutions and in communities. The scientific community is one among the moral communities of mankind and its unity is unintelligible apart from the commitment to realism. Thus the continuities in the history of that community are primarily continuities in its regulative ideals, and the realism which informs those ideals appeals to standards of truth and understanding which are pre-Tarskian (although not necessarily inconsistent with Tarski's theses). To state them adequately would involve the identification of a historical task, set in the sixteenth and seventeenth centuries and perhaps never to be completed. The building of a representation of nature is, in the modern world, a task analogous to the building of a cathedral in the medieval world or to the founding and construction of a city in the ancient world, tasks which might also turn out to be interminable.

To be objective, then, is to understand oneself as part of a community and one's work as part of a project and part of a history. The authority of this history and this project derives from the goods internal to the practice. Objectivity is a moral concept before it is a methodological concept, and the activities of natural science turn out to be a species of moral activity. It is therefore less surprising than it seemed at first that the history of the philosophy of science should replicate the history of moral philosophy.

It is also less surprising that the history of science itself cannot, after all, be written intelligibly solely as a succession of theories ordered by rational criteria. The history of science is a complex set of dramatic narratives in which individuals are at key points forced to gamble on future outcomes in just the way that the arguments of Pascal, Hume, and Nelson Goodman on induction would lead us to expect. (The wrongheadedness of all attempts to "justify" induction lies precisely in the attempt to show that this is not a gambler's universe, a kind of metaphysical poker game.) When Boltzmann committed suicide, partly at least, it appears, because of the growing scepticism among certain scientists about the kinetic theory, he behaved appropriately as a man who had staked his all on the realism of the kinetic theory, on the possibility of carrying through that moral task which he had made central to his own life. Science is a morality.

Nonetheless, there is at the same time a permanent, paradoxical and irremediable tension between science and ethics. In the Aristotelian world view, science and ethics were carefully guarded from conflict. The highest goal of the good life was the contemplation of scientific truth, and the realization of human potentiality in the achievement of the good took its place in the hierarchy of transition from potentiality to actuality which constitutes the natural world. Every species has its telos and the achievement of the telos by men includes the discovery of this unity of man with nature. The unity of that world view was of course precisely what rendered it vulnerable; for it continued to impose the concept of the unity and perfection of the cosmos as requirements upon a collection of proto-sciences that was, in the fourteenth and fifteenth centuries, heterogeneous and conflict-ridden. The science that replaced it retained the concept of unity, but made central to itself the project of describing the universe entirely in nonteleological terms, indeed in terms that would as far as possible be independent of the standpoint of any particular observer. Nature is to be represented as unitary and nonintentional.

The paradox that emerges is that the scientist can describe nature only in terms that exclude any characterization of science itself. For science is clearly an intentional activity; and here I refer not only to the activities of individual scientists, but to those projects and purposes which emerge gradually and both transcend and include many different generations. Yet, any description of nature, including human nature, in terms of what Weisskopf calls the quantum ladder will be one in which intentionality disappears from view. Hence, it would be absurd to expect moral conclusions to follow from scientific premises, not because of any general logical principle forbidding inferences from "is" premises to "ought" conclusions—there is no such principle—but because of the particular character of natural science.

This means that we inhabit a basically incoherent universe; nonetheless, we can give good reasons for giving morality primacy over science. For science can only become intelligible to us through its history and the continuities of this history are moral continuities. Our knowledge of nature is, we know, immensely fallible; the best confirmed and most adequately justified theory in the whole history of science did, after all, turn out to be false. But our knowledge of our knowledge of nature, that is, our knowledge of the history and philosophy of science, is far better founded. Yet its foundation depends in turn on our understanding of the history of human relationship to goods. Ethics and the history of ethics are not outlying territories after all.

Commentary

Response to Alasdair MacIntyre

Marjorie Grene

As ONE WHO COMES from such positively Martian territories as continental philosophy and other places uncharted by any philosophical map, I cannot judge MacIntyre's claim that the more "practical" branches of our profession are generally considered outlying in relation to metropolitan problems. The bewildering accumulation of work on "action theory" or the proliferation of Rawlsiana, to mention only two recent very noticeable explosions, might suggest the contrary. So too would the boom in medical ethics, of which the present remarks form a minuscule part. Be that as it may, if MacIntyre's thesis that "we can give good reasons for giving morality primacy over science" may be welcome to moral philosophers threatened by logical imperialism, it is at least as welcome to those of us interested in questions about the structure and significance of cognitive claims. For these are indeed, as MacIntyre argues, ingredients in human practices, which, like all practices, demand description and analysis in respect to the norms, traditions, and even ontologies inherent in their history and their structure. Thus the message of MacIntyre's paper for the philosophy of science and its relation to ethics, as well as for the historical understanding of science, is an extremely important one. What I am doing in these notes, therefore, is to raise a few minor questions about what I recognize as essentially a true, as well as a salutary, tale.

First, the title. The upshot of the paper *seems* to be: "Objectivity in science is moral." Yes, indeed. Gerhard Krüger said many years ago that even the *KrV* is grounded in at least one norm: *Wahrhaftigkeit*. To that extent there is a morality *of* science, even for Kant. But is MacIntyre also telling us something about objectivity in morality insofar as there is ethics as distinct from the ethics of science? I'm sure he will tell us in his forthcoming book; but meantime, is this a J.L. Austin kind of paper, that is, a paper with a title it doesn't exactly speak to, or am I just missing half its argument?

Moreover, the story MacIntyre tells us, though in general convincing, contains some surprising twists. For example, one does not expect Kierkegaard and Emerson to appear as leading representatives of nineteenth-century ethics, nor, for that matter, Kuhn and Feyerabend as leading representatives of twentieth-century philosophy of science. Further, one wants to ask, if what ethics scooped philosophy of science in was "the commission of *error*," was the orthodoxy against which these pairs of heretics were rebelling *truth*? In the case of Kierkegaard, who was, in Hegelian fashion, decrying Hegel, MacIntyre as a loyal Hegelian may in fact subscribe to the implicature suggested by his statement. What he says, however, is that Kierkegaard was quick to denounce Kantian ethics, a style of ethical theory which does not, I believe, represent for MacIntyre the truth in that field. Indeed, I don't really believe that Kierkegaard was much of a moral philosopher at all. It was the "aesthetic" life (which would have horrified Kant, had he dreamed of its existence) and the "religious" that engaged him. He was quick to see that the test for the categorical imperative failed, because ethics bored him in any case.

In the case of Emerson, I suppose the rebellion against a Unitarian father goes back, in negative-dialectical stages, to a rebellion against the moral theology set aside by Kant. Again, however, Kantian ethics does not appear, in the whole of MacIntyre's story, as the truth against which his heretics rebel, but rather (and I think rightly) as half of the uneasy tension between nature and morality that we have to live with. It was indeed Kant's oddly asymmetrical division of reality, with the moral will given a practical grasp of the noumenal, while poor old theory found itself afloat on a sea of pure phenomena—it was this division that gave birth on the one hand to philosophies like positivism (or logical empiricism) and to "null-point philosophies" like Sartrean existentialism on the other: antithetical styles of philosophizing, each of which tries, one if you like "scientistically" and the other pseudo-moralistically, to derive an account of all of experience from one half of the Kantian dichotomy (which is itself a demireduction of the Cartesian dichotomy of *res extensa* and *res cogitans*). In short it was not a true vision, but one already badly skewed by the critical philosophy, against which Kierkegaard and Emerson in their different, religious *bzw.* antireligious, ways, were rebelling: an exacerbation of error in each case, not a substitution of error for truth. I am not, by that statement, trying to contradict MacIntyre's narrative, but simply to stretch out a little an account so closely textured as to appear, so far, more knotted than it need.

Again, on the side of philosophy of science, MacIntyre's account is at times a little hard to follow. Look at the second half of his central thesis: that philosophy of science follows ethics in its commission of error, viz., "Kuhn's reincarnation of Kierkegaard and Feyerabend's revival of Emerson—not to mention the counterpart to them all, in which the eighteenth century outbids the nineteenth, Polanyi's version of Burke."

To put Polanyi center stage in the development of the philosophy of science is on principle a welcome move. Over twenty years ago, when he had published a brief version of his program for philosophy of science in the *Cambridge Journal* (7, 1954: 195-207) under the Hume-like title "On the Introduction of Science into Moral Subjects" (Cf. *A Treatise of Human Nature: Being an Attempt To Introduce the Experimental Method of Reasoning into Moral Subjects*), somebody (it may even have been MacIntyre, though I don't think so) suggested that what Polanyi was really doing was to reverse the direction of Hume's *Treatise* by introducing moral subjects into science. It seems that the need for some such reversal is at last coming to be recognized, although MacIntyre is one of the few (along with John Ziman in *Public Knowledge,* for example) who have recognized in the course of their own arguments the existence, let alone the importance, of Polanyi's work. Usually his work is either decried (as, for instance, almost compulsively by Colodny in the first few volumes of the Pittsburgh philosophy of science series), ignored (as in Richard Burian's recent paper on the relation between the history and philosophy of science, for example), or, in its more superficial aspects, quietly borrowed. Very few people indeed have troubled even to plagiarize the most significant aspect of Polanyi's philosophy, the theory of tacit knowing, which he stated best in the first chapter of *The Tacit Dimension* and in a few of the papers in the collection, *Knowing and Being*. There are, I am sure, numerous reasons why all this should have happened, including Polanyi's seemingly unphilosophical rhetoric, as well as the ill-conceived cosmology in which he tried to embed his account of science. But that is beside the point; the point here is simply that when I read MacIntyre's account of rules of practice and the way practices develop, together with the passages in which he describes and stresses the significance of Polanyi's model of science, I find myself (as one often is with MacIntyrean arguments) swept happily along to say, "Yes, how excellent and how true." When I look at the details of his paper, however, though still assenting enthusiastically to his analysis of the chief dimensions of Polanyi's characterization of science (its emphasis on the role of norms in science, its stress on tradition— including, I should add, a tradition of criticism; this is not, as was believed at Pittsburgh, a blind authoritarianism—its fideism and its intrinsic, more than semantic, realism) and though still assenting also to MacIntyre's own analysis of practices, including science, I am puzzled by some passages in his paper and even in a way by the relation of the second to the third part of his argument. I don't really know what he is saying about whom or what and why. One almost wants to remark that if Polanyi's rhetoric puts people off, MacIntyre's puts them on.

At least, the scenario we are given is alarmingly complicated. "Modern ethics," we are told, "has shown a striking tendency to anticipate the philosophy of science in the commission of error." So we have Kierkegaard to Kuhn, Emerson to Feyerabend, and "counterpart to them all," Polanyi=-Burke. But please: does this mean, as the paragraph says, that Polanyi has committed the gravest error of all? Then why does MacIntyre's own account of science sound so very Polanyian? And if (as he certainly was *politically*) Polanyi was another Burke, what has Burke to do with Kant? Where has our scenario gone? There was:

God

Inexorable *lex naturalis*: planets, which never disobey

Moral *lex naturalis*: people, who sometimes do disobey

Then there was: G̸o̸d̸

(or with Kant: God in the wings)

planets and other predictable natural processes or entities and their explanation by science

people as unpredictable beings not amenable to scientific explanation

ethics cut off from science

philosophy of science devoid of ethics

Next we have:

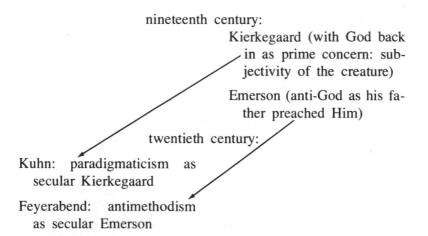

nineteenth century:

Kierkegaard (with God back in as prime concern: subjectivity of the creature)

Emerson (anti-God as his father preached Him)

twentieth century:

Kuhn: paradigmaticism as secular Kierkegaard

Feyerabend: antimethodism as secular Emerson

But then we get:

> eighteenth century:
>
> Burke: tradition as ruling
>   normativity of practice
>   realism (*how*, by the way, is Burke a "real-
>   ist?")

all this being (it seems to me) non-Kantian:

> Kant's ethical-political theory is pure
>   enlightenment, not traditionalist
> Kant's norms in *KpV* are formal,
>   not imbedded in practice
> Kant's science is phenomenal,
>   his ethics restricted in its realism

twentieth century:

Polanyi: science as normative
                    traditionalist
                    realistic

Kuhn: coarse stress on dis-   Feyerabend: rejection of all
continuity in the above   the above

What story *is* this? Feyerabend is presented as rebelling, in the last analysis, if via Popperian deviance from the original, against the "tyranny of the Vienna Circle." Certainly; and in a way, so was Kuhn. The Hypothetico-Deductive method was a fairy tale and one that kept attention away from the realities of science. And it was indeed, as MacIntyre insists, the purification of science from its roots in ethics, and hence in the norms of practice, and so from any recognizable connection with its own practice, and thus with itself, since it *is* a practice: it was indeed that separation of science from ethics that was largely responsible for making philosophy of science, from its inception until very recently, a philosophical wasteland. Polanyi, too, however, was trying to correct precisely the misinterpretation of science by scientism that Feyerabend rejects. How can Feyerabend be rebelling in the same move against the Vienna Circle *and* against

Polanyian science? It seems to me that he is rebelling against the aridity and unreality of a positivistic model of science while refusing, along with his empiricist cousins, to admit the normative, authoritarian, and realistic components of science as a practice that they all unite to decry. It is true that Polanyi, Kuhn, and Feyerabend are all fideists of a sort. Is Polanyi more in error than the others because his fideism is more global and systematic? By now I am quite bewildered. I feel as if I were watching a play by Tieck, perhaps *Der gestiefelte Ethiker* or *Die Verkehrte Wissenschaft*?

And then, to cap the climax, comes point three, which I find in the main entirely clear, convincing, and uncannily reminiscent of Polany's account of science, as well as of the hints given in *Sein und Zeit* (especially section 69b) of what an existential analysis of science might look like. True, MacIntyre emphasizes, in his report of Polanyi as well as of Polanyi's alleged derivative, Kuhn, the dimension of discontinuity between theories implicit in the parallel drawn by Polanyi between western naturalism and Zande belief in witchcraft. But Polanyi's account does also, I believe, take cognizance of the continuity in the growth of scientific knowledge much as MacIntyre describes it. Thus, MacIntyre's description of the way in which the term "acid" has changed its meaning as part of a continuous growth of chemical knowledge is reminiscent of Polanyi's account, for example, of the change of meaning in the term "isotope" or, say, of Rudwick's report of the change of meaning of the term "fossil," which I have myself recently used as an example of continuous growth, as distinct from revolutionary discontinuity, in science. Moreover, MacIntyre's description of the way in which internal goods relate (internally) to community seems to me very like Polanyi's account of intellectual passions as contrasted with appetite. Kuhn's relativism, in other words, appears to result rather from an overemphasis on one aspect of Polanyi's theory of science: the logical gap, both between evidence and theory and between one conceptual framework and another, than from a negligent or vulgar reading of the whole of what Polanyi has to say. Further still, MacIntyre's statement of the nature of objectivity, and of the fundamentally historical and *therefore* moral nature of science is entirely in harmony with the major bent of

Personal Knowledge. Finally, even the discord between science and ethics with which MacIntyre concludes corresponds closely to Polanyi's sense of that disharmony insofar as I understand both their views.

Nor do I question, any more than Polanyi would have done, MacIntyre's final remark about the fallibility of scientific knowledge. I do wonder, to return to my initial question, whether he is suggesting an objectivity in ethics more universal and apodictic than the fragile objectivity of science. That would sound (to me) more like Habermas than like Polanyi (who, after all, described the whole of our human heritage of beliefs in all areas as "all that in which we may be totally mistaken") and even less like what I, in all conscience (but of course fallibly) believe to be the case.

Let me repeat in conclusion, however, what I suggested at the start: What MacIntyre has presented in this paper is clearly an extremely condensed version of a much more intricate argument; an argument not only welcome, but necessary at this juncture in the history of our conception of science. I have been pointing to some obscurities which are due, I am sure, only to the demands of condensation on the present occasion.

2

The Moral Psychology of Science

Stephen Toulmin

"The Reason is, and ought to be, The Slave of the Passions"
—David Hume

I

In recent years, one significant component of earlier philosophical discussion—what used to be called "moral psychology"—has fallen into unjustified neglect. Yet, to the extent that *motives* are of crucial importance to ethics, the philosophical analysis of talk about "motivation" can play a helpful part in showing how values achieve practical expression in actual conduct; and this remains true even where, on the face of it, the activities in question are as "purely intellectual" as those of the natural sciences. So, in the present chapter, I shall be raising questions about the *personal engagement* in sciences of scientists as human individuals.

We have lived too long (I shall argue) with the image of the basic scientist as the solitary embodiment of the pure reason—"voyaging through strange seas of thought, alone," like some isolated hadron moving at high velocity through field-free space—and of the life of science as committed to a passionless objectivity. So, a myth has developed about the enterprise of natural science that too often deceives both its advocates and its critics. Supporters of science are tempted to take pride in its supposedly disinterested, dispassionate, and "value-free" character, in which they see the cutting edge of the objectivity that is its chief source of intellectual power.[1] Natural science is successful (on this view) just because its inquiries are purely "cognitive" and so undistorted by the affective preoccupations of individual scientists. Meanwhile, the opponents of science see in this purported objectivity the mark of its chief failing. Natural science ends by impoverishing our knowledge of the world (they retort) just because it shuts its eyes to all those affective channels through which alone real understanding might eventually be achieved.[2] I shall argue here that both these views are mistaken in their shared misconceptions about the "objectivity" of scientific work. Instead, we can hope to understand the values of science aright only if we first get our ideas about the motivation of scientific inquiry—and the moral significance of that motivation—back in proportion. Affect without cognition may be aimless; but cognition without affect is powerless, if not actually blind. The task is, therefore, not to *deny* the affective components in the life of science, but to understand them.

True, we have already begun to nibble away at the edges of the old image. For example, Frank Manuel's *Portrait of Isaac Newton* depicts the man who was for so long the totem figure of "rational" modern science as having worked under the impulsion of quite unexpected ambitions and confused self-perceptions.[3] Still, discrediting Newton's embodiment of the scientific ideal has not wholly reinforced the determination to "try harder." So, it is time to look again at the assumptions underlying the traditional ideal, in particular, at the notions that rationality itself can be a kind of motivation, and that activities pursued in a strictly "rational" spirit are to that extent "passionless."

If we accept these notions uncritically, we risk serious category confusions, which may lead to a needless undervaluing of the emotional (or affective) components in the life and work of science. This, in turn, may tempt us to play down both the individual moral choices involved when we commit ourselves to such a life, and also the moral accountability of the scientific community generally. By contrast, if we can develop a more satisfactory account of scientific motivation, this should bring with it a number of dividends. It should allow us to describe the collective scientific enterprise in more acceptable terms; to circumvent those implicit claims to moral immunity that have been—alternatively—both a major charm of the scientific life to those who pursue it, and a source of irritation to those who are only onlookers; and finally—not least—to begin extricating ourselves from the wreckage of those older dichotomies (cognitive/ affective, rational/emotional, thought/feeling, etc.) that still obstruct the road toward an adequate understand of those human enterprises in which feeling powers thought, and emotion makes rationality effective.

II

How far, then, can rationality itself serve as a "motive" for scientific work? The chief products of science certainly include a great many *arguments*; indeed, the very success of a science might even be measured by the number and quality of the new arguments being generated within it at the time in question. Consequently, scientists pay a lot of attention to the merits of scientific arguments: the relevance of new evidence, the power of novel concepts, the soundness of proposed inferences, and the weight these all give to the resulting conclusions. As a result, the working currency of a science is largely comprised in the intellectual power, weight, or force of its arguments.

This intellectual power, weight, or force has two distinct aspects. For practical purposes, working scientists often treat the merits and shortcomings of scientific arguments as something *inherent* or *intrinsic*. Either the available evidence really is sufficient to support some conclusion, or it is insufficient, or its significance is marginal; either some new concept really does have powerful implications, or it does nothing much to help us, or its merits are as yet unclear; either sound inferences do really force some conclusion on us, or they fail to do so, or judgment needs to be suspended. . . . In each case, all competent and informed scientists will presumably reach the same positions, given only a knowledgeable and accurate scrutiny of the arguments in debate. For what *has* intellectual weight (i.e., intrinsic intellectual merit) deserves to *carry* weight (i.e., to be persuasive); and a scientific apprenticeship will presumably give the experienced scientist an ability to recognize intrinsic merits when he sees them.

All the same: leaving aside all questions about "deserts," we should be clear that the intrinsic merits of an argument alone do not ensure its actual capacity to move a reader or hearer. The power, weight, or force of an argument (or its elements) is not in itself a "motive." Only those who are *open to* the argument in question will be influenced by it: in order to be persuaded, they must be ready to change their minds in response to an argument of this particular kind. To consider whether some argument has the actual power to persuade a hearer or reader is thus quite a different matter from considering whether it has "intrinsic" intellectual power. It is to raise the question of "power," "weight," or "force" on a different level and in a different sense.

Indeed, it is even to treat the notion of an "argument" itself on a different level and in a different sense. If regarded as a network of propositions whose interrelations can be judged in intellectual terms, an "argument" has merits that can be discussed without reference to the number or character of the human beings who would acknowledge those merits. If regarded as a clash of opinions which we may participate in, be carried away by, and win or lose (by contrast), an "argument" has merits that are directly related to its persuasiveness. It is of little use for the defense attorney or scientific presenter, political speechwriter or business adviser, to have "powerful" propositional arguments at his disposal, if his actual oral arguments are so feeble that they leave everyone unmoved.[4]

With this initial distinction in mind—between *arguments* (power, weight, force, etc.) as concerned with the intellectual relations between propositions, and *arguments* (power, weight, force, etc.) as concerned with the actual persuasiveness of those interconnected statements—we can at once make two further points. In the first place, (1) an argument is operative within the scientific enterprise only if it actually *weighs with* some actual scientist; only if "the" reasons it provides for accepting certain conclusions engage the concerns of some actual scientist, and so can become "his" reasons for subscribing to those conclusions. This does not always happen at once. For some thirty years, the arguments that Gregor Mendel had formulated about "inheritable factors" as the key to genetic transmission left other botanists and zoologists unmoved. Precisely because of this fact, we now see Mendel's effective contribution to biology as dating only from the restatement of those arguments by Bateson, Correns, Tschermak and de Vries around the year 1900. (Were Mendel's arguments really *powerful* before 1900, then? This question is ambiguous. How we answer it will depend both on how we interpret the term "powerful," and also on how seriously we take the theoretical debate about heredity that went on during the 1860s, when Mendel first put forward his theories.)[5]

In the second place, (2) although we can sometimes focus on the intrinsic merits of scientific arguments in isolation from all individual scientists, this fact does not entail that the actual enterprise of science is purely "cognitive," and so strictly rational, affectless, and value-free. By focusing our attention in this way, we too easily turn away from all questions about how the scientists involved, whether individually or collectively, *in fact came* to take the intellectual steps they did; instead, we end by asking only what steps it was *appropriate and proper* to take, given all the available considerations. Maybe they were, in fact, moved by undiluted curiosity, maybe by conformism or a desire to shock, maybe by a mixture of motives: it all depends on how far they were ready to be moved by the intrinsic merits of those particular novel arguments. And this, in turn, all depends on their having been both psychically disposed, and also intellectually prepared, to accept them.

To say that an individual scientist is *psychically* disposed to accept an argument is to say, among other things, that he is personally free to move that way—or, indeed, to move at all, rather than clinging to earlier opinions despite new arguments. To say that he is *intellectually* prepared to do so is to say, furthermore, that the move represents a straightforward and manageable change from his earlier position.

So the power, weight, and force of scientific concepts, arguments, and evidence are operative within the enterprise of science—can show their power by appearing compelling to scientists, and so "weighing" with them—only when they are capable of engaging the specific concerns of the scientists involved. In a phrase, the intrinsic merits of scientific arguments are manifested in the activities of actual scientists only when their *disciplinary relevance* enables them to do so. The "intrinsic power" of novel scientific arguments is matched by an equal "effective power"—their proper intellectual merits are translated into an operative capacity to persuade—only if they are directly relevant to the currently accepted concepts, theories, and problems of the science concerned.

III

This, however, is only a preliminary conclusion, and we need to pursue the point further. If scientists are to accept the arguments they do, it is not only necessary that they have the *particular* interests required; it is necessary also—and *a fortiori*—to recruit their *general* scientific interests. And it is here that the direct route into the individual moral psychology of science is customarily blocked off. Having reached this point, we are invited to assume that any particular scientist's commitment, whether to science in general, or to some specific line of inquiry, is a direct response to the intrinsic charms of intellectual worth. The "virtues" which Alasdair MacIntyre sees lying at the heart of the collective enterprises of science are presumed to have a self-evident attraction for particular scientists.[6] The fact that "good science" is capable of engaging the interests of individual human beings at all is taken as self-explanatory.

Yet, surely we need only make this assumption explicit to see how dubious it is. A taste for natural science or pure mathematics or legal theory is no more universal than a taste for classical music or pop art or modern dance; and the divergences of taste, even among committed scientists, are as great as those between (say) lovers of Vivaldi and devotees of Stockhausen. Once we are of a mind to give it our trained attention, the cogency of a mathematical proof or the elegance of a physiological experiment may constitute a "reason" for being moved by it; but however powerful that intrinsic "reason," it is not an automatic "motive" for accepting the resulting theorem or experimental result. For that, we also need to have appropriate "springs of action": we must, for instance, be susceptible to the formal fascination of mathematics, intrigued by the art of experimental demonstration, preoccupied with these specific physiological issues, or otherwise motivated to focus our attentions on the matter in hand.

Even after acknowledging the claims of disciplinary relevance, therefore, we are only making a first beginning on the individual psychology of scientific inquiry. Granted the significance of the professional collectivity in defining the virtues of any science and determining the effectiveness of its arguments, we must also consider how individual scientists come to join that collectivity in the first place. Since the inclination to take up the work of science is not a universal human characteristic, the decision to do so presumably involves a mobilization of impulses, inclinations, ambitions and/or affects, which might alternatively have been channeled in quite other directions. Curiosity and competitiveness, spontaneous joy and the wish to please, internal censorship and delight in play, grandiosity and openness. . . . all the components normally involved in the development of an individual's general personality, character, and mode of life can presumably find expression in the particular activities of his intellectual life; and indeed, without the intervention of some such "springs of action," his professional work would scarcely have got started. In short, the natural sciences exist at all only because there is something in the work of science *for* the individual scientists involved.

What sort of thing will that "something" be? In answering that question, we shall have to pay close attention, on the one hand, to the motives that inform *any* kind of fruitful and effective human collaboration. Collective institutions and enterprises of all kinds—churches and universities, guilds and brotherhoods—develop their own patterns of expectations and standards of achievements; and in this way they come to define their own characteristic "virtues." So, we come to know what it means, in general terms, to be a "devoted" priest or teacher, master craftsman, or whatever. On the other hand (we must recognize) the virtues characteristic of any collective enterprise must be capable of "running with the grain" of the individual persons involved. The activity of being a professional minister or physician, research scientist, or whatever must bring its own real psychic rewards to individual practitioners, if the collective enterprise itself is to survive.

What kinds of psychic rewards are in question? Once again, at this point, we risk an ambiguity. (The "virtues" of a scientific enterprise can be as equivocal as the "powers" of a scientific argument.) For those who are *already* committed to the life of science, the only indispensable "virtue" is embodied in *the doing of science* itself. All fully committed scientists, that is, derive satisfaction from the conscientious and effective pursuit of their scientific work. (There's a tautology for you!) To say no more than that, however, leaves untouched several deeper questions: (1) how the work of science *comes to have* charms for those who do eventually commit themselves to it; (2) why those charms are apparent only to some and invisible to many others; and (3) what other, ancillary considerations may work together (synergize) with this central "virtue" to retain and reinforce the interest of one or another particular scientist.

Once we start looking at particular individuals, indeed, we may even come to doubt whether the intrinsic merit of science as such—either the essential "virtue" of a scientific enterprise, or the inherent "rationality" of intellectual work itself—either *must be,* or ever *in fact is,* wholly its own reward. There are so many routes by which other more familiar impulses, passions, affects and the rest can be recruited, and so many other more familiar rewards to be gained from participation in this work, that the assumption of any specific and essential "motive" as the only spring of action "proper to" science can be called into question as multiplying entities *praeter necessitatem.*

Run through the individuals of our acquaintance who are committed to some given branch of science, asking in each case what deeper concerns feed into that individual's involvement in the science; we may not have to dig very far before we bring these to light, and we shall often end by giving different answers about each person. Just now and then scientists have made their extra-scientific motives explicit, as when Jacques Loeb explained how he had moved from philosophy into physiology in the hope of pursuing the problem of free will more effectively.[7] But the existence of such idiosyncratic motives can frequently be inferred only from circumstantial evidence: from the individual scientist's extracurricular activities, from the topics he chooses for commencement addresses in later life, from the style of his work or his writing, and so on. (One biologist's scientific interests may reflect his theological commitments, since in his eyes the marvels of evolution still testify to the Wisdom of the Creator; another may value evolution, rather, as an ally in discrediting the belief in a Divine Creator; a third may find in evolutionary biology a psychic support for our ethical beliefs; and so on.)

Without our assuming any single impulse specific to scientific work as such, we can thus recognize that many more general springs of action are capable of being mobilized in ways which give sufficient motivation to one individual or another to pursue the work (even, at times, the drudgery) of his particular science, and so to be open to the intrinsic power of its ideas and arguments. The mere existence of a sound argument, accordingly, no more guarantees a motive to accept its results than the mere existence of a duty guarantees a motive to perform it. Despite all of Kant's counsels of perfection, the Reason is no more completely dependent for its scientific fruitfulness on evoking a pure emotion of "respect for rational principles as such" than Duty is wholly dependent for its moral effectiveness on mobilizing an unadulterated feeling of "respect for the moral law."

Some individual scientists, mathematicians, and other scholars, of course, may finally achieve the status of "intellectual saints," having dissociated their intellectual activities from any last dilution by ulterior motives and come to pursue truth itself out of a "pure will." Even so, this "pure will" still presumably mobilizes and channels affective inclinations directed earlier toward more worldly goals. As we find it embodied in actual human beings, that is to say, the pure love of truth is a somewhat sophisticated and far from self-explanatory product of the individual's personal development. In that sense, the very existence of science depends on every active scientist having had a prior affective investment *of some kind* in the work of science; and in this respect we can go along accordingly with David Hume's view that "Reason" is capable of being efficacious only if it is, among other things, a "Slave of the Passions."

IV

Confronted with this demand for a developmental psychology of science that does justice to the affective springs of scientific interest and inquiry, some people will respond by questioning whether anything useful can be said about these springs of a truly *general* kind. No doubt, we can collect anecdotal life histories about particular individuals who in due course became scientists, and ask what specific motives drew them into scientific work. (Of all the descendants of Charles Darwin and T.H. Huxley in their generations, for instance, why did Julian Huxley and Richard Keynes follow their ancestor into biology, while John Maynard Keynes became an economist, Geoffrey Keynes a surgeon and literary critic, and Aldous Huxley a novelist?) But is there a good reason to suppose that we can establish any definite regularities— to say nothing of universal laws—governing the psychological paths by which some talented young people find their ways into science, rather than into other alternative vocations or professions?

This doubt is one that we must respect. The whole field of individual psychology is exposed to a standing temptation to overgeneralize; and it should be a sufficient objection to any psychological theory, if it leaves no scope for individual variation and idiosyncrasy. However, the need to respect the proper demands of individuality does not wholly eliminate the need for general theories or theoretical investigations. To the contrary, even if we acknowledge the importance of collecting detailed life histories about the psychic development of particular scientists, the problem, *in what terms* these life histories should be stated if they are to be fully illuminating, will still need to be faced. And, at this point, we can indicate just why the developmental affective psychology of science is a crucial problem for traditional theory. For it is precisely the continuing inclination of psychologists to discuss "cognition" and "affect" in isolation from one another that is the most serious obstacle to a fruitful attack on this problem.

Thus, on the one hand, developmental psychologists in the tradition of Piaget write about the sequence of tasks and achievements that a child masters in the course of "growing up," without paying any close attention to the specific impulses and/or emotions that serve as the child's "motivation" for attacking those tasks. (Somehow, it seems to Piaget that a detailed scrutiny of that sequence of tasks itself will reveal in due course sufficient *implicit motives* to account for the child's passage from each "stage" in its cognitive development to the next.)[8] Meanwhile, the classical psychoanalytic approach focuses on emotional conflicts in a child's life of kinds that can lead to functional disablement in later life, without considering directly why some functions and abilities may prove intrinsically more satisfying to the growing child than others. (Yet an understanding of the ways in which psychic development can go off the rails, emotionally speaking, leaves entirely unanswered all questions about *what it is* for a child to develop in emotionally healthy ways.) Only in the last few years has any serious sign appeared of ideas that may help to bridge this gap in the vocabulary of psychological theory.[9]

The starting point of the present paper was, thus, the observation that "reasons" and "motives" are not themselves two kinds of motives; on the contrary, other "springs of action" are needed if we are to pay attention to, be moved by, accept the conclusions of, or act in accordance with even the best of "reasons" or "arguments." As a result, we must be skeptical about all attempts to classify the activities of individual agents according to the supposed "rationality" or "nonrationality" of their *motives*. Such dichotomies as "cognition versus affect," "intellect versus emotion," and "reason versus passion," too easily become invitations to fall back into the initial elementary confusion of reasons with motives. Rational enterprises are not, as such, required to be free from passions of all kinds: they are simply enterprises within which we are expected to discount and set aside all *fruitless and irrelevant* passions.

On one natural interpretation of the term, indeed, a wholly "passion-free" enterprise could never even get started—hence, the force of the tag from Hume with which I began. (At the very least, an effective natural scientist must "have a bent for," and be "drawn toward," scientific inquiries.) Nor is it helpful to suppose that there are any purely "cognitive" activities, from which all elements of "affect" and/or "volition" have been banished. Once again, the question is not *whether or no* our affects/emotions/impulses and other springs of action are mobilized in the course of "cognitively structured" activities; rather, it is *how* these springs of action are recruited and put to work toward specific "cognitive" ends. Nor, for that matter, can we usefully depict the life of the intellect as being, in itself, intrinsically "unemotional." Where (we might ask) would such human activities as science and mathematics be without the joy of discovery, the anxiety of suspense, or the disheartenment of frustration?

The insistence on treating the "rational" and the "emotional" as essentially opposed often serves, in practice, as a rhetorical device for insuring that public, collective interests and considerations are given priority over private, personal ones. The term "rational" is then used simply as a label for collectively relevant (intersubjective) considerations, "emotional" for personally relevant (subjective) ones. Even without any dependence on the dubious contrast between "internal" feelings and "external" reasons, private or personal interests are presumed—on this view—to weigh with an agent by way of his emotions, whereas public or collective interests supposedly operate on him by way of his intellect. Yet, this way of presenting the matter misrepresents the contrast between the two kinds of interests. Public, collective interests may provide an agent with "good reasons" to act as he does; but he will be moved to act in conformity with them only if they also engage his personal concerns, and so mobilize his springs of action. Correspondingly, whatever considerations ultimately channel his conduct, and/or its justification—whether the resulting action is self-seeking or self-sacrificing—the agent may act either deliberately or impulsively, either coolly or in the heat of the moment. The vigor of his emotions, or the character of his affective involvement, thus has no direct bearing on the personal or collective nature of the interests served by his conduct. Collective interests can be pursued with passion, personal interests with calm deliberation.

Instead of purging our rational enterprises of affective content, we have in fact an alternative course open to us. William Gass has written perceptively about the ontogeny of artistic creation and aesthetic sensibility, and his line of thought can be extended for our purposes.[10] As they enter into the psychic development of individual human beings, Gass argues, art and aesthetic response typically involve a "stylization" of play and desire. However "formal" their aims and merits, works of art are sophisticated expressions of the same basic impulses and wishes that earlier led the child into less strictly organized activities; while the viewer or listener's aesthetic response to a work of art likewise recruits feelings and inclinations that operated during childhood in less focused directions. The philosophical problem about art, as a result, is neither to demonstrate that artistic and aesthetic impulses are absolutely unique, nor to equate—and so "reduce"—them to other, less exalted drives. Rather, it is to show how the progressive differentiation and "stylization" of their products or outcomes transforms the child's early impulses toward play and delight into their distinctive adult forms.

If art and aesthetics reflect the stylization of native impulses— e.g., play and desire—natural science similarly represents the channeling and directing of other equally native impulses: exploration, curiosity, or whatever. With science as with art, to focus attention on "ontogeny" is not to deny the special character of science, nor to equate, still less "reduce," scientific activity to infantile curiosity or childhood exploration. The "rationality" of natural science, too, can thus be seen as residing less in the specific motives of its practitioners than in the form and structure of its end products. The passions that move scientists to pursue their research in adult life will have moved them earlier, as children, to explore the world (or their own ideas) in less formally structured ways. When they eventually commit themselves to the life of the intellect, the demands of rationality simply transform and channel the directions in which those same "passions" operate: i.e., determine the outcome of activities whose motives and substance remain the same passions that have been with them since childhood. (More exactly, the demands of rationality begin to determine the manner in which those later outcomes are to be *publicly criticized and justified*.)

If anything is unique about science, it is thus the character of its collective procedures and results. There is no additional virtue in assuming that all those who become scientists are activated by a single specific motive directed solely at science. On the contrary, the central question in the psychology of science is how— given its abstract concepts, its demanding procedures, and its rigorous standards of criticism—the rational enterprise of science is able to engage and mobilize preexisting affective energies available in the personal development of its participants; and, in particular, to what special, typical vicissitudes this process of development may be exposed along the way.

V

Finally, then, supposing that we reopen these questions about the individual psychology of science, why should this make any difference to our ideas about the *ethics* of scientific work? (Why do I label the resulting insights as contributions to *moral* psychology?) This question, too, needs to be addressed on the collective, as well as on the personal level. We must ask both how the collective choices of scientific institutions and professional groups are to be judged from a moral point of view, and also how the individual scientist's commitment to science is to be subjected to moral scrutiny.

The road into the psychology of science has been blocked off (I argued) because, by polite convention, only one motive for doing science has been accepted as professionally respectable: viz., an unworldly love of truth and/or rationality for its own sake. There is, of course, a strong element of pious fraud to this convention. Yet, once we have accepted it, we become powerless to prevent the moral neutralization of the scientific enterprise. To the extent that the choices made by scientists in the course of their work are indeed made under this one comprehensive motive—as being so many alternative expressions of the scientist's "passion for rationality"—it might perhaps be legitimate to abstract the *intellectual content* of those choices wholly from their *moral context,* and judge them solely and exclusively in terms of disciplinary criteria and requirements. And this is something that many professional scientists are quite happy to see done.

Whatever helps to make the intellectual priorities of scientific inquiry a matter for criticism by disciplinary standards alone helps also to reinforce the professional autonomy of science, and so protects the collective scientific enterprise from interference by "outsiders." But this kind of autonomy can be achieved and maintained only at a certain *political* price. It may be professionally desirable to encourage single-minded concentration on the techniques and problems of a single scientific field among the younger apprentices to the professional guild of science. But it is no longer practicable today to shield the collective affairs of the scientific community entirely from outside criticism, including *ethical* criticism, than it is to protect (say) the professional institutions of law and medicine against the demand for public scrutiny and accountability. (It might throw dazzling light on the human brain if neurophysiologists decided to study "the influence on higher mental functioning of psychoactive agents administered in the form of airborne droplets," as being a ready and effective method of investigating the biochemical keys to various neural pathways; but that scientific decision will look very different if it is judged in the context of, say, a program of work on nerve gases sponsored by the East German Department of Defense.)

Similarly, on the individual plane, once we set aside the phantasmagoria of "rationality for rationality's sake," which merely isolates our motives for intellectual work from all other intelligible motives, moral issues arise once again about any particular scientist's involvement in his work. Suppose, for instance, that a scientist is entirely candid about the charms that his work has for him. He will, no doubt, say a good deal about how he was first drawn to the concerns of science, particularly *his* science—how his early childhood curiosity became initially engaged with wildlife, and in due course focused in on (say) flying beetles; or how he came to value the delayed gratifications or professional fame and craftsmanlike self-satisfaction as substitutes for earlier and more direct rewards, such as parental approval. But this universal, commonplace scenario is unlikely to tell the whole story, and he may also acknowledge other, more mixed and even questionable motives. For instance, by devoting his personal energies to prestigious, but highly abstract inquiries, he has been able to avoid emotionally demanding situations in which he would have had to deal with complex and conflicting personal demands directed at him by others. And the impulse to avoid such external demands may prove, in turn, on closer self-examination, to spring less from proper pride than from fear of failure. In all honesty, therefore, he may finally conclude that his own commitment to the life of science has, on a certain level, been psychically self-serving, and even "inauthentic." His devotion to (say) entomology, while sincere enough in itself, has been strengthened and reinforced by other, more personal psychic rewards.[11]

So, when we ask ourselves what there is *in* Science *for* scientists, it is not enough to answer—whether ingenuously or disingenuously—"Why, Truth and Rationality, of course." Rather, we should accept this question in all its richness and acknowledge the full psychic and moral complexities of the motives that can lead us, as private persons, to find individual satisfaction as a result of committing our professional lives to (say) topology, or physical cosmology, or psychopharmacology. In particular, it is an intriguing question of psychodynamics, how talented scientists learn to distance themselves from their own ideas, in the interest of scientific criticism, and reconcile themselves to the apparent sacrifice of any personal "psychic interests" in their own brainchildren.

Is this process facilitated (for instance) by early apprenticeship to a charismatic teacher, whose standards the student "internalizes," so recruiting all the forces of conscience, paternal-filial attachment, and/or the "superego"? Must this process be reinforced by the collective institutions of professional life, and so by a sense that any falling away from the "rational ideals" of science will be an infraction against the scientific collectivity, as well as against the individual's conscience? Or is it normally strengthened, in addition, by an admixture of narcissistic grandiosity, i.e., by the individual scientist's tendency to impose unrealistic standards of achievement on his own performances, as a defense against the fears aroused by his own human fallibility? Will there thus, from a personal point of view, be a common element of perfectionism and overscrupulosity in this insistence of exemplifying in one's own person the undiluted devotion to truth and rationality conventionally claimed for the collective scientific enterprise?[12] These questions by themselves hint at the complex issues of personality development and moral judgment that need to be addressed in this field; and the scanty material currently available by way of evidence leaves us in no position to do more than speculate about possible answers.

VI

For the purposes of moral psychology, the key term in our discussion is perhaps "vicissitudes." To the extent that the affective development of future scientists displays any typical episodes, leading to the development of recognizable "psychopathologies," the nature of scientific motivation becomes a deeper and more genuinely *ethical* issue. Thus, Gerald Holton recently raised questions about the ethical concerns of young scientists, along lines hinted at above.[13] Faced with emotional conflicts (he suggests) many talented young people today display a certain psychic frailty: a commitment to the somewhat abstract inquiries of natural science can then give them a psychic "defense" against the pain of having to deal publicly with complex ethical issues. The single-valued nature of the traditional scientific life thus shields them from the emotional difficulties of other more worldly pursuits, and in this way they acquire a vested *personal* interest in the claim that the natural sciences are "value-free."

This hypothesis can easily be misunderstood. In suggesting that ethical frailty or psychic defensiveness plays a part in the recruitment of some scientists, Holton is not of course putting forward any universal generalization; still less is he implying—like Theodore Roszak and other recent romantic critics of science—that a wholehearted commitment to modern science is in itself the mark of ethical insensitivity. (On the contrary, his evidence suggests that it is, if anything, the mark of ethical *hyper*sensitivity.) So it may be helpful to compare Holton's thesis, at this point, with another claim that many people have also found perplexing: viz., Freud's thesis that Art is a sublimation of Sex—i.e., that artistic creation and sexual life are alternate adult expressions of the same "libidinal" energies, affects, and/or impulses, which are present in us all from infancy.

If Freud is right, the question may always be raised about any individual painter or writer or musician,

> How completely have the artistic and sexual aspects of libidinal expression become differentiated in the course of his development? How far does his commitment to the artist's work rest on a single-minded devotion to the demands of the art, as such; and how far is it still colored by an admixture of artistically irrelevant motives, whether sexual or others?

(To pose Freud's question in these terms is not to imply that single-minded purity of commitment, uncolored by other motives, is even desirable, to say nothing of indispensable. In Art as much as in Science, mixed motives may well be the order of the day.) If Holton is right, parallel questions may always be raised about any individual scientist:

> How far is his commitment to science personally disinterested, and directed solely at the specific virtues of the particular scientific discipline in question? And how completely is the charm that science's abstract inquiries have for him free from any suspicion (say) of an admixture with moral frailty or other scientifically irrelevant motives?

This may seem at first glance a topic for individual psychology alone, but the issues involved cannot be wholly detached from their collective context. For the established virtues of any collective enterprise must (as I remarked earlier) "run with the grain of" the personalities of the particular individuals who make up the collective. So, the professional standards of the community of science may, in certain respects, reflect not just the methodological needs imposed on scientific work by the problematics of scientific disciplines, but also the common personal needs and failings typical of "scientists" as human individuals. And this observation may serve, in conclusion, to throw some light on the deeper roots of the analogy that John Ziman has drawn, between the professional life of twentieth-century science and the life of the medieval orders.[14]

It is not just that the scientific life is in many respects a

twentieth-century successor to the medieval monastic life. To go
further, the "monastic orders" of modern science have largely
become orders of *contemplatives,* whose devotion to Truth excludes other more worldly obligations. Now, one recurring question in the literature of monasticism is an ethical question: viz.,
how an order of pure contemplatives can remain in a state of
grace. It is not that the rigorous separation of contemplatives
from the world exposes them to an increased temptation toward
worldly pleasures. Quite the contrary: the contemplative mode of
life erects barriers between them and the world, which too easily
acquire a charm of their own and become an excuse for moral
irresponsibility.

The professional isolation of the modern scientific life may
well give rise to similar problems. The disinterested pursuit of
scientific truth does not enhance the attraction of scientists toward
other more worldly ends. Rather, the "purity" of the scientific
life has its own powerful attractions, which easily seem to justify
a turning away from ethical issues. So, once we are launched
into an examination of the moral psychology of science, in both
its individual and collective aspects, we may find that the *vita
scientifica* involves ethical difficulties and complexities no less
acute than those traditionally associated with the *vita monastica,*
which was its prototype and forerunner.

NOTES

1. The image of "scientific objectivity" of the *cutting edge* of discovery is used by writers whose views of science may differ quite
   substantially in other respects, although the general tone of their
   arguments is inclined to be positivistic. For a useful historical
   treatment, see Charles C. Gillispie, *The Edge of Objectivity* (Princeton, N.J.: Princeton University Press, 1960).
2. This is the burden of the argument in, e.g., Theodore Roszak's
   book, *Where the Wasteland Ends* (Garden City: Doubleday, 1972).
   See also my discussion of the antiscience movement in the CIBA
   Symposium, *Civilization and Science* (Amsterdam and New York:
   Associated Scientific Publishers, 1972).
3. Frank Manuel, *A Portrait of Isaac Newton* (Cambridge, Mass.:
   Harvard University Press, 1968), assembles a remarkable body of

argument about Newton's deeper motivation: specifically about the Divine "calling" he felt to decipher the cosmic handiwork.

4. For a fuller discussion of this point see my book, *Knowing and Acting* (New York: Macmillan, 1975), esp. Part II.

5. The immediate fate of Mendel's ideas has been frequently discussed in recent years. For a useful analysis, see E.B. Gasking, "Why Was Mendel's Work Ignored?", *Journal of the History of Ideas*, vol. 20, pp. 60-84, 1959.

6. See Alasdair MacIntyre, "Objectivity in Morality and Objectivity in Science," in this volume.

7. Jacques Loeb, *The Mechanistic Conception of Life* (Chicago: University of Chicago Press, 1912; reprinted, Cambridge, Mass.: Harvard University Press, 1964).

8. For a fuller discussion, see T. Mischel, Academic Press, ed., *Cognitive Psychology and Epistemology* (New York: 1969); and C.F. Feldman and S.E. Toulmin "Logic and the Theory of Mind," *Nebraska Symposium on Motivation, 1975* (Lincoln, Neb.: University of Nebraska Press, 1977).

9. On this topic, I have personally learned a great deal from discussions at the Center for Psychosocial Studies, Chicago, and at the Chicago Psychoanalytic Institute.

10. See William Gass, *Fiction and the Figures of Life* (New York: Knopf, 1970).

11. Compare Lionel Trilling's interesting series of essays, *Sincerity and Authenticity* (Cambridge, Mass.: Harvard University Press, 1972).

12. For relevant discussions of narcissism, grandiosity, infallibility and associated phenomena in the development of emotional life, see H. Kohut, *The Analysis of the Self* (New York: International Universities Press, 1971) and *The Restoration of the Self* (New York: International Universities Press, 1977).

13. See Gerald Holton, "Scientific Optimism and Societal Concerns," *Hastings Center Report*, vol. 5, no. 6, December 1975, pp. 39-47.

14. John Ziman, *Public Knowledge* (Cambridge, Eng.: Cambridge University Press, 1978).

Commentary

Response to Stephen Toulmin

Gunther S. Stent

I AM VERY MUCH IN SYMPATHY with what I take to be the main point of Stephen Toulmin's essay: to understand science it is necessary to take into account that it is practiced by flesh-and-blood persons influenced by affects, and not by disembodied spirits or intelligent automata operating solely by pure reason. As indicated by Toulmin's lead quotation from Hume, such spirits or automata would be useless, since "reason is, and ought to be, the Slave of the Passions." However, in contrast to Toulmin, I believe that this truth happens to be one of the few philosophical points regarding science of which most working scientist are, in fact, aware. Indeed, even the authors of science fiction movies seem to know this: the automata "Hal" in *2001* and "R2D2" in *Star Wars* are clearly driven by affects in the performance of their intellectual feats, malign in the case of "Hal" and benign in the case of "R2D2." Where is Toulmin going to find supporters of science that are still tempted to take pride in its supposedly disinterested, dispassionate, and value-free character, in which they see the cutting edge of the objectivity that is its chief source of intellectual power? Not among the participants in the current public controversies regarding sociobiology and genetic engineering, to take two examples. Here the main force of both the attack against, and the defense by, the practitioners of these disciplines is directed toward personal motives rather than the attainment of objective truths. The sociobiologists' and genetic engineers' main defense rests, not on the grounds of rationality for rationality's sake, but on the benefits that their noble, selfless, and devoted labors will bestow on mankind. On the other hand, sociobiologists and genetic engineers are usually attacked, not on the grounds that they shut their eyes to "affective channels," but that they are guided by the wrong affects; they are accused of being racists, reactionaries, or mercenaries, or whoring for a Nobel Prize.

Thus, one of my general criticisms of Toulmin's essay is that it does not sufficiently connect its thesis with real, everyday science. First, although Toulmin hints that epistemological consequences flow from Hume's dictum, he does not trouble to present any case histories in which "moral psychology" can be clearly identified as a component of what a substantial group of contemporary scientists hold to be an objective truth. It would have been easy to find an example from sociobiology, but just how is moral psychology reflected in some specific, currently accepted proposition of cosmology (e.g., big bang), high-energy physics (e.g., quarks), molecular biology (e.g., genetic code), or neurophysiology (e.g., split brain)?

Second, Toulmin claims that heuristic consequences flow from Hume's dictum, in that the passions enter into the force of scientific arguments. By way of an example he cites a single case, namely, the failure of Mendel's discovery of the gene in 1863 to move his fellow biologists for the next thirty years. Unfortunately, Toulmin's discussion of this famous paradigmatic case of delayed recognition is not only brief and quite unclear, but Toulmin even declares that its relevance to the central point at issue is ambiguous. It would have been good if Toulmin *could* have shown how passion delayed acceptance of Mendel's discovery of the gene, precisely because here the link between heuristics and moral psychology is far from obvious. [By contrast, the link to moral psychology of the religious controversy about Darwin's natural selection theory (which Toulmin does not cite), which raged during the very eclipse of Mendel's discovery, is so obvious as to render that example too trivial for making any deep point.]

Third, Toulmin claims that moral consequences flow from Hume's dictum, in that, contrary to what "many professional scientists are quite happy to see done," the intellectual content of the choices made by scientists cannot be divorced from their moral content. Here he provides as an example the study of the influence on higher mental functioning of psychoactive drugs, administered in the form of airborne droplets, and points out that although this method might be an effective one for the study of brain biochemistry, the decision to make that study "will look very different" in the context of a nerve gas research project on behalf of the "East German Department of Defense." But this example is not very cogent, since research on the effect of psychoactive drugs is not considered free of moral content even if conducted in the Free World under the sponsorship of the United States Department of Health and Welfare, as anyone trying to do such research would immediately learn from the forms needed for approval of such a project. To make his point, Toulmin should have taken on the rather more illuminating, real (rather than hypothetical) case of scientists deciding to accept sponsorship in the 1950s and 1960s (prior to the prohibition of this practice by the "Mansfield Amendment") by the United States Air Force or Navy of esoteric research projects utterly remote from any conceivable military objective. Or, better yet, Toulmin might have considered the even harder (and ethically more significant) case of the moral implications of Darwin's natural selection theory, which, according to some Asian and Third World scientists, follow from the fact that it came about as a result of a project sponsored by the Royal Navy in the service of British imperialism.

Another general criticism I would make of Toulmin's essay is that its foray into the psychology of motivation rarely rises above the statement of commonplaces. For instance, by way of accounting for why certain persons become scientists, rather than choosing some other line of work—a choice of which it would certainly be good to have an account—Toulmin says that there must be "something" in the work of science *for* that person. And, as Toulmin explains, neither the love of rationality for rationality's sake nor the "virtue" inherent in being a scientist is likely to be the only, or even foremost, component of that "something." But what that "something" is, or how we might proceed to discover it, Toulmin does not say. It would be unfair, of course, to ask of Toulmin to provide for the special case of the scientist a solution of the general problem posed by Freud's fundamental insight that the true motives for our actions are usually not those of which we are consciously aware. After all, that problem has baffled generations of analytical psychologists, who have yet to develop reliable procedures for ferreting out "authentic" motives from the infinitude of a person's psycho-historical facts. But to Toulmin's insight, that the genesis of the scientist's superego can be accounted for in terms of an internal-ization of standards derived from paternal-filial and teacher-stu-dent relations, reinforced by the institutions of his environment, a psychoanalyst is likely to respond: "So, what else is new?"

Indeed the decisive importance of personal psychohistory for understanding the judgments of individual scientists is itself dem-onstrated by the evident differences between my perception and Toulmin's of the generally held moral attitudes vis à vis modern science. The scientific community may very well be regarded as a monastic order of Newtonian contemplatives at the Oxbridge high tables, which, I imagine, formed Toulmin's outlook. But I find it hard to believe that this view is widely held elsewhere.

I received my primary education in Nazi Germany, whose scientists were included among the "workers of the forehead," who, jointly with the "workers of the fist," followed the Fuehrer's lead in building the Thousand Year Reich. In turn, the refugee scientist initiators of the Manhattan Project felt it their duty to see to it that Hitler's Reich would not outlast a thousand days. No room for monasticism there. Later, I received my higher education in the Chemistry Department of the University of Illinois, under the banner of "Better Living Through Chemis-try," the slogan of the DuPont Company, whose research and management was then mainly in the hands of Illinois alumni. Here the inventor of nylon, and not the contemplator of universal truths, was the role model. Finally, I spent my entire academic career on the Berkeley campus of the University of California, whose laboratories have long been dominated physically, if not intellectually, by the "Hill," or Radiation Laboratory of the Atomic Energy Commission. So, in view of this background, I did not meet many scientists who have been taken in by what Toulmin calls the "pious fraud" of claiming that the only respec-table motive for doing science is an unworldly love of truth.

Nevertheless, I heartily endorse Toulmin's call for a project on the "Moral Psychology of Science." Only, I wish that in this essay Toulmin could have gone further in showing why this project is needed, what it would do for us if it did achieve its aims, and how to go about doing it.

3

The Origins of Psychoanalysis Revisited: Reflections and Consequences

Steven Marcus

ALTHOUGH I CHOSE as my general topic "The Relation of Psychoanalysis to Morals" or "The Influence of Psychoanalysis on Moral Values," I should like here to confine myself to a more modest and manageable scope. I should like to examine some of the scientific, medical, cultural, and personal circumstances out of which this new discipline arose. And I should like to sketch briefly how it arose out of them. That this nexus was highly charged with moral or ethical implications and that what emerged from them has bearings upon our large or general conceptions of what constitutes moral life is, I should suggest, not a matter for argument.

Since I am a literary critic, I have chosen to focus my attention on a text. That text consists of the letters, notes, and drafts that Freud wrote and sent to Wilhelm Fliess between 1887 and 1902. It is an incomplete text because of the 284 surviving documents only 168 were published in *The Origins of Psycho-Analysis,* which first appeared in German in 1950 and in English in 1954.[1] It is further incomplete in the sense that significant parts of some of the 168 published documents have been edited out (some of these parts can be recovered in Ernest Jones's[2] biography and in Max Schur's *Freud: Living and Dying*).[3] And it is incomplete as well in the sense that no parts of Fliess's side of this intense correspondence and lengthy relationship in writing seem to have survived. Nevertheless, most texts are incomplete in one sense or another. All texts are constructions, and the student of literature soon becomes accustomed to the awareness that one of the central problematics of his discipline has to do with what it is that constitutes a text. Once one has put aside the accepted conventions about what a text is, a theoretical snake pit opens up. I will not at this point enter such a snake pit, but will content myself with observing that just as there is hardly a text worth considering that does not present us with problems—and problems about its own theoretical status—so too there is nothing to deter us from treating a set of letters and related documents, even an edited set, as a text as well. As long as we preserve the integrity of such a text as we have (and here I am departing from certain currents of advanced contemporary critical practice) we are no more constrained than we are when we deal with a play by Shakespeare, a novel by Dickens, an autobiography by, say, John Stuart Mill or Thomas Carlyle, or a journal by Gerard Manley Hopkins. So I think that without more ado we can turn to the text in question.

The letters that Sigmund Freud wrote to Wilhelm Fliess between 1887 and 1902, along with the notes, drafts, and other material that Freud included with those letters, make up an extraordinary account. They constitute a singular record of creative discovery in Western thought, and are indispensable to our understanding of how psychoanalysis—itself a unique development of Western civilization—was brought into being. That development may be appropriately regarded as a culmination of the particular tradition of introspection which began with the adjuration of the oracle at Delphi to "Know thyself." This rationally governed method of self-examination takes as its principal objects of scrutiny everything within us that is not rational—our affects, our instinctual strivings, our fears, fancies, dreams, and nightmares, our guilts, our endless reproachfulness, our sexual obsessions, our uncontrollable aggressions.[4] And it is more than a fortunate accident that this most highly developed form of Western secular introspection should have returned at one of its moments of climactic breaking-through to its cultural origins. When Freud, in October of 1897, writes that "I have found love of the mother and jealousy of the father in my own case too, and now believe it to be a general phenomenon of early childhood,"[5] and then goes on at once to discuss *Oedipus Rex,* one senses reverberations of profound historical depth. An organic line of cultural evolution is being brought to a decisive moment of conclusion. That conclusion represents the attainment of a new degree of consciousness in Western civilization. Yet this new or higher degree of consciousness resumes what has gone before, incorporates and reintegrates its past and intervening history, and by thus returning to its original point of departure simultaneously moves forward into a future that has been significantly altered.[6]

The Freud we meet in these documents is the person within whom this historical drama of development first occurred. At the outset he was certainly not consciously prepared for the direction in which such a development would lead him. He was a young neurologist or neuropathologist with a generous range of interests and high ambitions. He had already published scientific papers on a variety of topics—including the use of cocaine—and had in Paris come under the influence of Charcot, whose notions about hysteria and sexuality he had accepted. He was also interested in the new discoveries that were taking place in the study of the anatomy of the brain and the nervous system, and in such related pathologies as aphasia and the battery of afflictions that accompanied syphilis in its advanced states. One large connection between these pathologies and the neuroses, in particular hysteria, was that they tended to mimic each other. Hysterically distorted speech or hysterical loss of speech functions often was indistinguishable from the reduced or deformed speech functions of persons who had, from one cause or another, incurred damage of the brain. And vice versa. The blindnesses, paralyses, tics, lamenesses, spasms, pains, and phantom sensations of hysteria were all produced as well by syphilis in its tertiary phase. Indeed the nineteenth-century epidemics of venereal disease and hysteria were connected in a number of intimate ways. Not only was their symptomatology often convertible and interchangeable, but they often tended to occur coincidentally or in tandem, and in the same persons. The cultural and moral history of these entities still remains to be written.

The other conditions with which neurologists largely occupied themselves—apart from what today are regarded as actual neurological illnesses—were obsessions, phobias, depressions of certain magnitudes, paranoia in its milder forms, and a variety of symptoms that were bundled together to form the syndromes of neurasthenia and anxiety neurosis. These conditions had a number of attributes in common—apart from the circumstance that no one understood them. One such attribute was essentially negative: they were all in some sense residua, left behind or overlooked by German psychiatry, which at that time and point in its historical development was almost exclusively concerned with psychoses and the other more florid and spectacular conditions of mental disorder. Such conditions all seemed, in addition, sooner or later to have something to do with sexuality, but this common feature too was understood, if that is the right word, in a primitive and unformed sense. The store of therapeutic measures that was available to the young Freud was equally rudimentary. The traditional pharmacopia was of very limited use. Other current resources or treatments of choice were electrogalvanic sessions, hydropathic or water cures, and rest cures of differing descriptions. Hypnotism was relied on by some, and Freud experimented with it, as he did with just about everything else. In addition to hypnotism accompanied by suggestion, there was also therapy by suggestion alone, and Freud seemed at the outset to rely on these two procedures as much as on any others. As he wryly observes to Fliess in one of the early letters, "Talking people into and out of things . . . is what my work consists of." Occasionally some of these methods worked, but no one knew why. More often they failed to make any difference, but again it was generally beyond anyone's power of explanation to say why a particular treatment didn't work in particular cases. As Freud's remark suggests, it was by no means a satisfactory state of affairs—neither therapeutically nor theoretically.

It was by no means a satisfactory state of affairs culturally either. Situated as he was, Freud could not help but observe and be struck by what appears to have been the plague that had settled upon the sexual lives of the middle classes in the latter part of the nineteenth century. The early letters to Fliess, like other medical writings about sexuality of the same period, and like much contemporary literature—such as the novels of Hardy and Gissing or the plays of Ibsen—all refract the same sexual climate of gloom, frustration, fear, muted despair, and moral uncertainty and bafflement. For example, quite early on, Freud connects neurasthenic impotence in men with a history of mastur- bation, and then connects in turn hysteria in women with neu- rasthenic sexual inadequacy in men: if worn-out men excite but cannot satisfy women, hysteria is the result. It is not quite to the point to say that Freud was mistaken; the value of these early formulations lies in what they suggest to us about the material that Freud was observing. In a similar way Freud was at first concerned with the adverse effects that *coitus interruptus* seemed to have on both middle-class men and women, and collaterally the even wider adverse effects that the absence of any innocuous form of contraception seemed to threaten. "In the absence of such a solution," he wrote at the beginning of 1893, "society seems doomed to fall a victim to incurable neuroses which reduce the enjoyment of life to a minimum, destroy the marriage relation and bring hereditary ruin on the whole coming generation."[7] And in the same document he declares that although neuroses can be prevented, they are at the same time "completely incurable."[8] Although he was soon to change his mind on this topic, from the very beginning Freud made and continued to make a firm connec- tion between his patients' range of distresses and states of distress in their sexuality.[9] Although he was at first able to explain very little about either the etiology or the dynamics of their unhappy circumstances, he was also unable to overlook what he had observed or had strongly inferred: namely, that the neuroses in all their differences were all connected in some profound but as yet undisclosed way by disturbances of sexuality. Having made this observation and inference he clung to the connections they pointed to with what would later prove to be characteristic tenac- ity. Others at the time had made similar connections. Yet, the question remained: what could result from these connections; to what did they lead; how was one to proceed?

Freud begins by trying to describe and classify the various

afflicted states that come before his attention, and tries at first to find a simple, physical causal explanation for the various neuroses. For example, in 1893, he tried to explain anxiety attacks as a direct result of sexual experience, adducing the case "of a gay old bachelor, who denies himself nothing." This patient, Freud continued, produced "a classic attack after allowing his thirty-year-old mistress to induce him to have coitus three times. I have come to the opinion that anxiety is to be connected, not with a mental, but with a physical consequence of sexual abuse."[10] And indeed, Freud's first psychological explanation of anxiety as either dammed-up libido or as a transformation of it may be properly regarded as a derivative of this still-earlier view. These physical or physiological theories had to do with notions of stimulus, excitation, building up of hydraulic pressures or electrical charges of sexual substances, and with such other notions as discharge, equilibrium, homeostasis, constancy, and the like. One sees Freud in these letters and drafts beginning again and again with such ideas and being led inexorably away from them by the complexity of the accompanying psychic states that he encountered. This is to say that the physiological theories that were available to Freud were simply and finally inadequate to deal with the richness of the psychic material produced by his patients. Freud did not move willingly toward the establishment of an independent psychology. He moved slowly and reluctantly, and one of the outstanding achievements of this early period, his long unpublished "Project for a Scientific Psychology" of 1895, is clearly a work of genius.[11] It is an effort to construct a unified systematic theory of the mind, rising in an unbroken set of connected stages from neurones to complex psychic states; it is not less than an effort to solve or transcend the mind-body duality, and it was, of course, abandoned. Nevertheless, after having given up this heroic undertaking and turned toward the slow and laborious construction of an autonomous psychology, Freud never forgot that the mind was in fact in the body; furthermore, by making questions of sexuality central to his psychology he made it certain that his chances of ever transcending this overriding duality remained minimal. And indeed one can say that in no other psychology is the mind-body relation so inclusively and centrally a presence as in Freud's.

Since its first publication in 1950, the "Project" has come in for a good deal of comment and considerable praise from such students of modern neuroscience as Pribram, and perhaps a further word on it might at this point be appropriate. The theory is based on the neurones, which had been discovered in 1891, and the model it follows is taken from neuroanatomy. It posits at first two systems of neurones, permeable and impermeable, and a series of "contact barriers" that separate and join them (the synapse had not yet been described). It has to do with such notions as energy, inertia, and homeostasis, and in particular focuses upon Q, or the electrical energy that builds up in and innervates the nervous system. Hence Freud's later notion of "psychic energy" has a clear physicalist base or origin. The energy in question is biological, or electrical energy as it exists in a biological organism. It is not directly mechanical or Newtonian energy. And indeed, the not quite unambiguous use to which Freud puts this notion of energy permits him to avoid the crude dualism of conceiving of the human body on the one hand as mere mechanism or machinery and the human mind or person on the other as conscious, willing, and goal-directed.

Yet when Freud abandons the "Project" he does not return to some simple Cartesian dualism either—to a dualism, say, of psychology and neurology. What he does is to retain the neurophysiological "hydraulic" model of energy, resistance, discharge, inertia, storage, and so on—but what he gives up is the *neuroanatomical* model, and the concomitant effort to locate such physiological processes in determinate anatomical positions in the central nervous system. Instead of that he substitutes a virtually or conceptually spatial "psychic apparatus" which acts as a kind of functional isomorph of the abandoned neuroanatomical model; this "psychic apparatus" provided him with a means of making functional localizations of psychic processes without committing him to discovering specific parts of the brain wherein this apparatus was localized. Yet he never at the same time gave up on the primary model except, as he said, "for the present." He believed that some day—in the neuroscience of the twenty-third century perhaps—the systems of his psychic apparatus would, in fact, in some as yet unimagined way, be localized.

I have always found something quite moving in this complex

set of strategic maneuvers managed by Freud. He remained a materialist without going over to some kind of positivism or systematic reductionism. He developed an autonomous psychology without committing himself either to idealism or to some form of Cartesian dualism. He remained loyal to the idea of science—even as the science of the late nineteenth century thought of itself—and his loyalty has, I believe, been rewarded. If I ask myself why I am so moved by these complex and sometimes obscure conceptual shiftings-about, I am prompted to reply that Freud's intellectual commitment and adherence to the idea of science has a profoundly moral component to it.

This vexed and difficult set of passages may be momentarily summed up as follows. Freud did not, on the one hand, make a reduction of the mental to the physical; and he did not, on the other hand, use these physical notions simply and solely as convenient explanatory models for mental functions. Nor did he merely trim his sails and settle for some compromise in between. It was the inadequacy of the physiological theories then available to Freud that moved him toward the creation of an autonomous psychology; he would never have taken the position that all physiological accounts of mental functioning were, by definition, bound to be inadequate. His belief in science was far too strong for that. At the same time, even though "the mind was in fact in the body," Freud did not describe the mind as a physical system, though he did use "energic" terms to describe the workings of the psychic apparatus.

If I may be permitted to use an analogy from literature, Freud seems to me, in this very large and general connection, to have demonstrated his powers of what Keats describes as "negative capability." "Brown and Dilke walked with me and back from the Christmas pantomine. I had not a dispute but disquisition, with Dilke on various subjects; several things dovetailed in my mind, and at once it stuck me what quality went to form a Man of Achievement, especially in Literature, and which Shakespeare possessed so enormously—I mean *Negative Capability,* that is, when a man is capable of being in uncertainties, mysteries, doubts, without any irritable reaching after fact and reason— Coleridge, for instance, would let go by a fine verisimilitude caught from the penetralium of mystery, from being incapable of remaining content with half-knowledge.''[12]

Freud's correspondent and interlocutor was admirably qualified to be the witness of such a complex journey of discovery. Wilhelm Fliess was also fascinated by the mind-body relation, and made its mysteries the center of all his researches and theories. With his hypotheses of nasal reflex neuroses and of the arithmetical periodicities that governed the course of all human— and even nonhuman—existence, Fliess was precisely the right kind of person on whom Freud could securely try out his new insights, theoretical flights, and systematic wild hunches. The careful reader of the documents I have been discussing can hardly escape the realization that for many of the years covered by these documents, Freud was something of a scientific wild man, a "conquistador or adventurer" as he described his own temperament. Great creative geniuses often are, and it was a necessity of Freud's nature that he needed a male companion to accompany him on his forays into the darker regions of mental existence. His friendship and professional alliance with Josef Breuer had come to an end on just these grounds; Breuer was unable to go along with Freud's embarrassing insistence upon the central importance of sexuality. Fliess took Breuer's place as Freud's confidante and alter ego. He was much younger than Breuer and two years younger than Freud. He was quite unembarrassed by considerations of sexuality; in fact he led Freud to place the notion of the bisexuality of all human beings close to the center of his theoretical system. Fliess was a scientific wild man as well; he was not, however, a genius, and his theories were not transmuted under the influence of new insight and experience, as Freud's were, and thereby made to conform more adequately to the differentially relative requirements of science, medicine, or psychology. Fliess's theories remained wild, were elaborated in ever more flamboyant displays of explanatory epicycles, and ended up as a kind of monomania. The occasion that provided the impetus for the final breakup of this most important friendship of Freud's life was, appropriately, a profound theoretical difference, which in this instance, as in others, had certain moral bearings; Fliess insisted upon believing that the arithmetical and numerological laws of periodicity governed everything, including psychopathology. Such a belief was clearly incompatible with Freud's developing notions of psychic determinism—notions that asserted whatever might *ultimately* be true about the physical laws that govern all phenomena, in our present state of knowledge mental existence has certain autonomous functions, and most mental activities can be traced no farther than to antecedent determinations that are themselves psychic.[13] When Fliess remarked accusingly of Freud (and to him) that "the thought-reader [Freud] reads only his own thoughts in those of others," the time to break had come. It was an appropriate pretext to end a relation that had been going downhill for some time. Freud parted company with Fliess when his friend was no longer able adequately to sustain the mind-body duality, and when he insisted upon depriving one side of that relation of its due importance and complexity.

Freud's friendship with Fliess was close, intense, richly personal as well as intellectual and deeply intimate. Though the two men met at irregular intervals, at what Freud liked to call "congresses," the relation was sustained in large measure through their correspondence. Of the 284 letters written by Freud to Fliess that have survived, all but forty were written in the years 1893-1900, the period of highest intensity in their friendship. Looking back upon that friendship from the perspective that Freud's own discoveries has enabled us to achieve, we can perceive that it was in the main constituted out of transference-like characteristics. Freud overvalued and idealized the intelligence, originality, and powers of judgment of his friend. He consulted Fliess about his own health, and submitted to Fliess's medical advice as if his ear, nose, and throat friend in Berlin were Aesculapius himself. He kept his own critical faculties in abeyance as far as Fliess was concerned. He looked to him for approval, and he dreamed about him—some of the dreams are included in *The Interpretation of Dreams*. Naturally and unavoidably, certain neurotic processes were involved in this relationship, and among the many things these letters permit us to observe are the points of connection between such processes and Freud's creativity during this period. They also allow us to observe the gradual disentanglement of the two. Freud's relation with Fliess was an essential and enabling part of that creativity. His resolution of that relation brings this major phase of his career to a close. With the publication in 1900 of *The Interpretation of Dreams,* Freud had reached a new plateau, an intellectual elevation from which he could regard large parts of the human world afresh.

The path he followed to reach that goal was not direct but circuitous, as these documents amply demonstrate. Freud was only gradually to be convinced of the primary independence of psychic reality. He began with materialist views and physicalist assumptions, and he was in no hurry to abandon them. The first stage in that evolution, as we have already seen, was represented by the "Project" of 1895—and by the fact that Freud gave it up. The second, and more dramatic, stage occurred in 1897. In working with his neurotic patients, Freud kept continually coming across accounts produced by them that pointed to certain sexual experiences in childhood, in particular among his female patients with accounts that pointed to experiences of seduction by their fathers. Freud took these accounts at face value, and proceeded thereupon to construct a theory of neuroses which had at its genetic center the actual experience of sexual seduction in childhood; at one of its dynamic centers was the idea that such patients were unable adequately to react to, or "abreact," the traumatic excitement of those experiences. This incapacity to somehow achieve a "discharge" of such highly stimulated states, even though such states remained behind as mental representations such as memories, fantasies, and dreams, was itself based on a physical model, and indeed the same states in question converted themselves into the bodily symptoms of hysteria. In September, 1897, Freud wrote to Fliess that he had been compelled to give up his theory that he had been building up for some years. The letter merits quoting at length:

> Let me tell you straight away the great secret . . . I no longer believe in my *neurotica.* That is hardly intelligible without an explanation; you yourself found what I told you credible. So I shall start at the beginning and tell you the whole story of how the reasons for rejecting it arose. The first group of factors were the

continual disappointment of my attempts to bring my analyses to a
real conclusion, the running away of people who for a time had
seemed my most favourably inclined patients, the lack of the
complete success on which I had counted, and the possibility of
explaining my partial successes in other familiar ways. Then there
was the astonishing thing that in every case . . . blame was laid
on perverse acts by the father, and realization of the unexpected
frequency of hysteria, in every case of which the same thing
applied, though it was hardly credible that perverted acts against
children were so general. . . . Thirdly, there was the definite
realization that there is no "indication of reality" in the unconscious, so that it is impossible to distinguish between truth and
emotionally-charged fiction. (This leaves open the possible explanation that sexual phantasy regularly makes use of the theme of
the parents.)[14]

Freud's response to this reversal, to the overthrow of a theory to
which he had given so much of himself and on which he had
placed such weight, is as remarkable as the reversal itself. "Were
I depressed, jaded, unclear in my mind," he writes, "such doubts
might be taken for signs of weakness. But as I am in just the
opposite state, I must acknowledge them to be the result of
honest and effective intellectual labor, and I am proud that after
penetrating so far I am still capable of such criticism. Can these
doubts be only an episode on the way to further knowledge?"[15]
And toward the end of the letter, he adds, "In the general
collapse only the psychology has retained its value. . . . It is a
pity one cannot live on dream-interpretation. . . ."[16]

It is characteristic of Freud's personal and intellectual courage—as well as of the ethical aspect of that courage—that he
should have regarded such a defeat as a partial victory and that
he should appear altogether undeterred by it. As almost any
reader of these documents will soon perceive, the years
1893-1900 are the *anni mirabili* of Freud's life and career. During
this period, living in relative isolation, Freud was literally a man
possessed and driven. He had found, he wrote, what he needed
in order to live—"a consuming passion—in Schiller's words a
tyrant. I have found my tyrant, and in his service I know no
limits. My tyrant is psychology; it has always been my distant,

beckoning goal and now, since I have hit on the neuroses, it has come so much the nearer."[17] The Freud of these years is possessed by a creative demon. He describes the demands he is making on his own mind as "superhuman," and speaks of psychology as "an incubus." When he writes to Fliess in December, 1895, that "we cannot do without men with the courage to think new things before they can prove them," he is describing himself.[18] Ideas, insights, partial theories were coming to Freud faster than he could handle them, set them down, or reflect upon them. Several months after he had written the "Project" he could write to Fliess that "I no longer understand the state of mind in which I concocted the psychology."[19] There is no reason to doubt the candor or truthfulness of this observation. For a large part of this period, Freud appears to have inhabited states of creative possession; his possession; his ambitions were of a magnitude that, fortunately for him, bore some plausible correspondence with the extent of his powers. On New Year's day, 1897, he writes: "Give me another ten years and I shall finish the neuroses and the new psychology. . . . When I am not afraid I can take on all the devils in hell. . . ."[20] The documents that make up the text I have been discussing are the firsthand record of how Freud took on those devils, most of which arose from the hell that was within himself as well as his patients. In May, 1897, he says: "No matter what I start with, I always find myself back again with the neuroses and the psychical apparatus. It is not because of indifference to personal or other matters that I never write about anything else. Inside me there is a seething ferment, and I am only waiting for the next surge forward."[21] He speaks regularly of things "fermenting" in him and of the "turbulence of my thoughts and feelings." Like a great poet, he had during this period exceptional access to his own unconscious, and parts of the turbulence he refers to has to do with the unremitting efforts he had to make to bring structure, organization, and meaning to this mysterious, dark, and threatening material. When he turns to the composition of *The Interpretation of Dreams,* he enters a state of creative transport and withdrawal. Having worked over a draft of one section, he writes to Fliess, "it is nearly finished, was written as if in a dream, and certainly not in a form fit for publication." Three weeks later, having finished work upon another section of the book, he sent it to Fliess with the following comments:

> It was all written by the unconscious, on the principle of Itzig, the Sunday horseman. "Itzig, where are you going?" "Don't ask me, ask the horse!" At the beginning of a paragraph I never knew where I should end up. It was not written to be read, of course—any attempt at style was abandoned after the first two pages. In spite of all that I of course believe in the results. I have not the slightest idea yet what form the contents will finally take.[22]

And Freud's son, Ernst, later recalled how during this period his father used to come into meals, from the place where he had been writing, "as if he were sleepwalking." Freud's own comments on this material—almost certainly the great seventh chapter, "On the Psychology of the Dream-Processes"—including his slightly benumbed incomprehension, suggest something of the depths at which he had been seized. Even after *The Interpretation of Dreams* had been finished and published, he still could write: "The big problems are still unsettled. It is an intellectual hell, layer upon layer of it, with everything fitfully gleaming and pulsating; and the outline of Lucifer-Amor coming into sight at the darkest centre."[23] If we are led in the context created by such allusions to think of Dante, Milton, Goethe, and Blake, we are not bringing in material that is irrelevant. The creative discoveries of Freud during these years were indeed of the highest order.

Those discoveries were legion. There is scarcely a notion that Freud was to elaborate upon in the course of his long subsequent career that is not touched upon, hinted at, presented in abbreviated, elliptical, or nuclear form in these letters and drafts. The writing itself is on occasion obscure and even incomprehensible, for Freud is moving about in darkness. Nevertheless, the main features of the mental terrain that he is exploring gradually come into focus and clarity. With an increasingly sure sense of intellectual priorities Freud concentrates his attention upon areas of major discovery: first, the meaning of dreams and the construction of a psychology of unconscious mental functioning; second, the occurrence and development of infantile and childhood sexuality; and third, the central place in that development occupied by the Oedipal experience. The major vehicle through which these discoveries were made was also an unprecedented occurrence. I am referring of course to Freud's self-analysis, a process in which, as I have already suggested, the classical instruction to know thyself was brought into historical and momentous conjunction with *Oedipus Rex*. That this process also fulfilled in part the quizzical injunction, "Physician, heal thyself," is only one further demonstration of the centrality of Freud's discoveries to the history and traditions of Western civilization. It also serves to remind us that the Freud who appears at the end of this correspondence is not entirely the same man who wrote the bulk of it. He is still, to be sure, a great genius and is on his way to becoming a great man, but he is not the wildly driven, creatively tormented, fitfully inspired, and demonic creature of the largest part of this text. Although Freud is a central figure in the European tradition of Romanticism, and one of its last important legatees, there is a classical side to his achievements and character as well. This side comes into view when we observe that the making of this genius coincided with the maturing of it. He was not afraid that the rational understanding of his own nature would undo the creative energies that had sustained him. And he was equally unafraid that the achievement of personal integration was incompatible with the continuation of a creative and useful life.

The expression, "Physician, heal thyself," occurs within and is part of the dominant religious tradition of our civilization. Freud belongs to that group of writers and thinkers who during the last two hundred years have undertaken the major task of transforming that tradition by secularizing it. Freud himself was not religious in either temperament or sympathy. Nevertheless, like the other great modern figures he retained one belief, albeit in secular form, that belongs to that tradition. This is the belief in meaning—a belief of extensive moral and cultural consequence. In the instance of Freud it was a belief that all human thought and behavior had a meaning and meanings—that they were understandable, purposeful, had structure, and rose to significance.[24] For example, in 1897 he writes: "Perverted actions are always alike, always have a meaning, and are based on a pattern which can be understood."[25] And in the same year, he also states: "Phantasies arise from an unconscious combination of things experienced and heard, constructed for particular purposes. These purposes aim at making inaccessible the memory from which symptoms have been generated." Discussing the distortions that are built into fantasies that masquerade as memories, he remarks: "I am learning the rules which govern the formation of these structures, and reasons why they are stronger than real memories."[26] And he follows his discovery of "love of the mother and jealousy of the father," with the following observations: "If that is the case, the gripping power of *Oedipus Rex,* in spite of all the rational objections to the inexorable fate that the story presupposes, becomes intelligible, and one can understand why later fate dramas were such failures. Our feelings rise against any arbitrary, individual fate . . . but the Greek myth seizes on a compulsion which everyone recognizes because he has felt traces of it in himself. Every member of the audience was once a budding Oedipus in phantasy, and this dream-fulfillment played out in reality causes everyone to recoil in horror, with the full measure of repression which separates his infantile from his present state."[27] This deep belief in the meaningfulness of human experience and the equally deep belief in the power of the human mind to effect coherent explanations of that experience acted as general motives in the work of Freud. Without divine sanction and without a theodicy, the story of each human life could yet be rendered into some measure of intelligibility. There might even be moments of surpassing clarity. Such a moment occurred on January 16, 1898, when Freud wrote: "Happiness is the deferred fulfillment of a prehistoric wish. That is why wealth brings so little happiness; money is not an infantile wish."[28] If there is such a thing as secular wisdom, such a statement, it seems to me, approximates the form it is likely to take. The paths by which such knowledge was attained are imperishably recorded in the text I have been briefly examining.

That last quotation is, it seems to me, a piece of wisdom and belongs in a formal sense to the formal genre known as Wisdom Literature. That genre begins in our tradition with Proverbs and includes such writers familiar to Freud as Epictetus, Marcus Aurelius, Bacon, and Goethe. It is useful to remind ourselves how generously Freud's writings are studded with such remarks and observations. "Wisdom is the principal thing, therefore get wisdom: and with all thy getting get understanding" might easily serve as one of several epigraphs to part of Freud's achievement. What I have been trying to demonstrate is how, in the working out of that achievement, the scientific and intellectual discoveries made by Freud were inseparable from the moral values that informed such a work of discovery and the moral insights that emerged from it. Whatever one may want to say about other topics—such as the influence of psychoanalysis on our thinking about sexuality or our notions of childhood and child-rearing or the moral nature of the psychoanalytic situation itself—in the text I have been discussing there is no such thing as an autonomous or detachable ethical sphere.[29]

At various points in his later career Freud went out of his way to deny that he was a genius. What he had achieved in the creation of this discipline, therapy, made of inquiry and institution, he insisted, was largely a matter of character. One of the things he meant by this not entirely perspicuous remark is that he was not in the formative years of discovery afraid to confront unpleasant ideas, live in isolation, and stand squarely in "The Opposition"—as he was fond of putting it. But I also think he meant that psychoanalysis did not simply spring, as a system, out of his mind, out of contemplation or pure research. He meant, I think, that much of what he had discovered had been tested in his own life and was the result of his experience. Indeed, Freud's writings are on one side properly considered as part of "the literature of experience." Experience goes with great learning, which Freud wore very lightly; it goes as well with shrewdness in the analysis of character; and both are combined with brevity of phrase and irony. Such writing surprises the reader with a sense of its truth and simultaneously leads him to self-examination. And this brings me back to where I began. The notion of self-examination—and the value attached thereto—is surely somewhere close to the center of whatever it is we think psychoanalysis may mean.

NOTES

1. Sigmund Freud, *The Origins of Psycho-Analysis*, ed. M. Bonaparte, A. Freud, E. Kris, trans. E. Mosbacher and J. Strachey (New York: Basic Books, 1954). *Aus den Anfängen der Psychoanalyze* (London: Imago Publishing Co., 1950).
2. Ernest Jones, *The Life and Works of Sigmund Freud*, 3 vols. (New York: Basic Books, 1953).
3. Max Schur, *Freud: Living and Dying*, (New York: International Universities Press, 1972).
4. Freud assumed that human existence is intelligible and that a coherent representation of all the psychic forces that are typical of human beings might some day be possible. Within the context of this presumptively rational model, certain behaviors or kinds of mental processes—namely, our affects, our instinctual strivings, etc.—are characterized as irrational.
5. Sigmund Freud, *The Origins of Psycho-Analysis*, Letter of 15.10.97, p. 223; this letter appears in *The Standard Edition of the Complete Psychological Works of Sigmund Freud*, trans. James Strachey, 24 vols. (London: Hogarth Press, 1966-74) as vol. 1, Letter 71, p. 265. All quotations are from *The Origins of Psycho-Analysis*. Reference will also be given to the corresponding passage in the *Standard Edition*.
6. The organic line of cultural evolution I am referring to has to do with the development in Western civilization of a tradition of rationally directed introspection or self-examination. Other cultures have other methods and traditions of introspection, and indeed Western civilization contains within itself alternative introspective traditions as well—*The Spiritual Exercises* of St. Ignatius Loyola, for example, are in some measure part of such an alternative tradition. The new or higher degree of consciousness I refer to has to do: (1) with the training of the rationally governed self-examination on the irrational and the unconscious; and (2) with the accompanying awareness, in the instance of Freud, that he was, in fact, completing a piece of work that had, historically, begun with the Greeks.
7. Sigmund Freud, *The Origins of Psycho-Analysis*, "Draft B: The Aetiology of the Neuroses," p. 72; *Standard Edition*, vol. 1, p. 184.
8. Sigmund Freud, *The Origins of Psycho-Analysis*, p. 71; *Standard Edition*, vol. 1, p. 183.
9. It is my impression that very few patients came to Freud because

they were generally "unhappy": The complaints tended to be fairly specific, if not multiple, and included the somatic symptoms of hysteria, the uncontrollable behavior of obsessional neuroses, frigidity, impotence, premature ejaculation, and various forms of perverse sexual gratifications. In the historical context of the period, such distresses all tended to be regarded as medical problems, although there was still a strong cultural component of moral condemnation when it came to perverse sexual behavior. In 1905, when he published *Three Essays on the Theory of Sexuality*, Freud did not include homosexuality among either the perversions or the neuroses. He put homosexuality into the separate category of "inversion" and decisively refused to reprehend it on moral grounds. He had not, by that date, ever had a homosexual as an analytic patient.

10. Sigmund Freud, *The Origins of Psycho-Analysis*, Letter of 17.11.93, pp. 79-80; this letter does not appear in the *Standard Edition*.

11. Sigmund Freud, *The Origins of Psycho-Analysis*, pp. 347-445, *Standard Edition*, vol. 1, pp. 283-397.

12. John Keats, *The Letters of John Keats*, ed. H. E. Rollins, vol. 1 (Cambridge, Mass.: Harvard University Press, 1958), Letter of 21.12.17, no. 45, p. 193.

In a thoughtful and useful essay, "Psychoanalysis, Physics and the Mind-Body Problem," (forthcoming in the *Yearbook of the Chicago Institute of Psychoanalysis*, Chicago: The Chicago Institute of Psychoanalysis, 1978), Stephen Toulmin discusses how under the influence of Kant, German scientists and philosophers had "authority, in a way that their colleagues in other countries did not, to justify their ignoring the more rigid dichotomies of *mind* and *matter*." Among those so influenced were such figures as Müller, Helmholtz, Meinert, and Wernicke, all of whom in turn were there to influence, in different degrees, the thinking of the younger Freud.

13. By "psychic determinism" Freud meant a number of things. He could mean, as he did here, that psychic processes constituted an autonomous domain. He could also mean that there are no psychic events without causes, and that when it comes to mental events there are no such things as accidents. He could also mean that there were no mental events without meaning. And he could mean as well that mental events are overdetermined. On occasion he combined several of these meanings, but it is usually not very difficult to tell which meaning he is enlisting by consulting the context in which the usage occurs.

14. Sigmund Freud, *The Origins of Psycho-Analysis*, Letter of 21.9.97, pp. 215-16; *Standard Edition*, vol. 1, Letter 69, pp. 259-60.
15. Ibid., pp. 216-17; p. 260.
16. Ibid., p. 218; omitted in the *Standard Edition*.
17. Sigmund Freud, *The Origins of Psycho-Analysis*, Letter of 25.5.95, p. 119; omitted in the *Standard Edition*.
18. Sigmund Freud, *The Origins of Psycho-Analysis*, Letter of 8.12.95, p. 137; omitted in the *Standard Edition*.
19. Sigmund Freud, *The Origins of Psycho-Analysis*, Letter of 29.11.95, p. 134; omitted in the *Standard Edition*.
20. Sigmund Freud, *The Origins of Psycho-Analysis*, Letter of 3.1.97, p. 183; omitted in the *Standard Edition*.
21. Sigmund Freud, *The Origins of Psycho-Analysis*, Letter of 16.5.97, p. 200; omitted in the *Standard Edition*.
22. Sigmund Freud, *The Origins of Psycho-Analysis*, Letter of 7.7.98, p. 258; omitted in the *Standard Edition*.
23. Sigmund Freud, *The Origins of Psycho-Analysis*, Letter of 10.7.1900, p. 323; omitted in the *Standard Edition*.
24. Freud did appear to believe that this search for meaning was a kind of functional norm; that it was a valid function; and that thoughts and behavior are expressions of that functioning.
25. Sigmund Freud, *The Origins of Psycho-Analysis*, Letter of 24.1.97, p. 189; *Standard Edition*, vol. 1, Letter 57, p. 243.
26. Sigmund Freud, *The Origins of Psycho-Analysis*, Letter of 7.7.97, p. 212; *Standard Edition*, vol. 1, Letter 66, p. 258.
27. Sigmund Freud, *The Origins of Psycho-Analysis*, Letter of 15.10.97, pp. 223-24; *Standard Edition*, vol. 1, Letter 71, p. 265.
28. Sigmund Freud, *The Origins of Psycho-Analysis*, Letter of 16.1.98, p. 244; omitted in the *Standard Edition*.

4

Reconciling Freud's *Scientific Project* and Psychoanalysis

Joseph Margolis

WE SPEAK EASILY of Freud's immense influence. And we qualify the concession, knowledgeably, by drawing attention to early resistance to Freud's ideas, the differences between the now-popular image of Freud and Freud's actual work, the general neglect of Freudian conceptions in areas in which one might have supposed Freud would have been carefully studied—for instance, the philosophy of mind,[1] doubts about the validity of putative contributions of Freud's, and some uncertainty about our own ability to say anything more instructive about Freud's actual distinction. It is perhaps useful, therefore, to begin with an overstatement. Freud's importance lies in his contribution to the theory of persons and cultures, though, in a perfectly clear sense, he has no theory of persons and cultures. Better still, he is concerned with selected aspects of the psychological dynamics of persons and cultures, without actually formulating what it is to be a person or a culture. The clarification of this distinctive contribution bears decisively, as we shall see, on the compatibility of science and the foundations of ethics.

Inevitably, there have been doctrinal schisms between Freud and his associates and his disciples and students. And there have also been a large number of quarrels about the conceptual adequacy and even coherence of Freud's larger theories. For example, regarding Freud's *Scientific Project*,[2] Karl Pribram has asserted: "I found that the *Project* contains a detailed neurological model which is, by today's standards, sophisticated. . . ."[3] On the other hand, regarding the *Project* and Freud's metapsychology in general, Irving Thalberg maintains that Freud's partition plans [that is, the id-ego-superego theory] are ". . . rooted in clinical and everyday experience. I am afraid they do not work. His consistently nonanthropomorphic explanations make sense, but fail to elucidate those aspects of reflexive and other behavior which most baffle us . . . [W]ith regard to Freud's animistic accounts, . . . these provide us illumination . . . [b]ut careful assessment will show these explanations to be instructively incoherent."[4]

Part of the dissatisfaction in all such quarrels lies with the sense of their partiality, of disputants' having selected fractions of Freud's thought. But a deeper part of their lack of satisfaction depends on the strong intuitive sense that the inaccuracies, the apparent incoherencies, the ingenuous errors, the pamphleteering, the piecemeal adjustments, the period prejudices, even the exuberant inventions of Freud somehow do not quite threaten the large strength of his endeavor. One searches, therefore, for a fresh perception of his fascination that is not to be denied because of the rejection of the oedipal complex or penis envy, the correction of the account of *The Madonna and St. Anne*, the disclosure of Freud's dependence on *The Golden Bough*, worries about his authoritarian character, and quarrels with his associates.

There is a clue in the *Project*, however, that also runs through the explicitly psychoanalytic material, that is of the largest consequence for the theory of science, in particular, for the program of science—the so-called unity of science[5]—that seeks to subsume the explanation of human behavior and of the behavior of human societies in terms adequate for the physical and the biological sciences. In order to appreciate the point, we must first consider a number of distinctions on which it rests.

Let us sort out, therefore, a number of senses of "emergence"—which bear on the *Project*'s clue. There is a sense of emergence, introduced by Paul Meehl and Wilfrid Sellars[6] and popularized by Herbert Feigl,[7] that maintains that the emergent "entail[s] a breach in *physical*$_2$ determinism," where, by "physical$_2$," Feigl intends "those concepts and laws sufficient for the explanation of inorganic as well as of biological phenomena" if "the type of concepts and laws which suffice in principle for the explanation and prediction of inorganic processes" may be extended to cover organic life.[8] In this sense, the admission of the emergent entails the denial of physicalism and of the program of the unity of science—and is opposed to reductionism in that precise sense. An alternative conception of emergence is proposed by Mario Bunge, who maintains that "an emergent thing (or just *emergent*) is one possessing properties that none of its components possesses."[9] Bunge's conception is qualified in a number of ways. For one thing, he wishes to relativize the concept of emergence. So he says, for instance, "the ability to think is an emergent property of the primate brain relative to its component neurons, but it is a resultant property of the primate because it is possessed by one of the latter's components, namely its brain." The example is quarrelsome, since there is a straightforward sense in which thinking is ascribable to organisms, agents, persons, and not to any of their "parts."[10] Nevertheless, Bunge's distinction is a useful one. Bunge also wishes, however, to defend at the same time at least the following four postulates:

> Postulate 1. Some of the properties of every system are emergent.
> Postulate 2. Every emergent property of a system can be explained in terms of properties of its components and of the couplings amongst these.
> Postulate 3. Every thing belongs to some level of other.
> Postulate 4. Every complex thing belonging to a given level has self-assembled from things of the preceding level.

Now, these postulates are intended to clarify two theses: the first, that reductionist explanation in the sense of the "physical$_2$" is precluded, since, on Bunge's view of emergence, the explanation of emergent phenomena "is a matter not of deducing consequences from a theory concerning some lower level, but of suitably enriching the latter with new assumptions and data"; the second, that minds "do not constitute a supraorganic level because they form no level at all": a mind, on Bunge's view, is not a thing at all, "but a collection of functions of neuron assemblies, that individual neurons do not possess."

Feigl's notion of emergence is open to challenge, as he himself very well saw, simply because it is not in the least clear that physical$_2$ laws can explain the psychological phenomena of the life of persons. Bunge holds as much as well. But Bunge's notion of emergence is open to challenge, in a way he does not fully appreciate, because although minds may not be things, persons are in some obvious sense "things"—things that, for reflexive reasons at least, *we* cannot disregard.[11] If we take seriously the emergent concept of a person, we see that it is at least problematic whether Postulates 4 and 2 can be correct. For example, on P. F. Strawson's view, whatever its shortcomings, persons are treated as "basic particulars," hence in a way in which their bodies or their bodily parts cannot be construed as *components* out of which they are "self-assembled."[12] But if that is conceded (or at least proposed as an as yet undefeated candidate), then Postulate 4 must be false or at least insufficiently supported. In short, Feigl's and Bunge's notions of emergence pretty well oblige us to concede a further notion, namely, one having a sense that entails Feigl's sense but is incompatible with Bunge's, a sense in which persons are emergent entities of some sort (not entailing Cartesian dualism), and cultures constitute the contexts in which they are identified and produce their characteristic activities and artifacts. This sense of emergence entails the rejection of the unity of science program in that respect in which Feigl's adherence to the adequacy of physical$_2$ concepts and Bunge's adherence to a compositional alternative both attempt to retain the purpose at least of the original program. This third sense of emergence holds, then, (1) the denial of the adequacy of physical$_2$ concepts; (2) the denial that persons are simply composed out of the elements of some lower level physical system; and (3) the denial that the explanation of the properties of persons—that is, the salient, culturally significant properties—can be explained in terms of the properties of the components of some physical or biological system. This is neither to deny that the behavior of persons and other cultural phenomena can be explained nor to affirm a form of Cartesian dualism. It is only to concede the coherence of a theory of persons and of culture that is not reductive in either Feigl's or Bunge's sense. The full details of this third alternative need not detain us here: it is only its coherence and potential usefulness that need be acknowledged. But what we may say is that Feigl's program is based on an identity thesis that precludes emergence; that Bunge's is based on a compositional thesis that admits emergent properties of composites; and that the third program admits emergent properties but replaces identity and composition with another relationship (precluding ontic dualism).[13] What invites the third program, very briefly put, is the emergence of linguistic ability and behavior informed by such ability.

We are now prepared for the distinction of Freud's endeavor. The *Scientific Project* and the flowering of psychoanalysis itself constitute, in a sense, a dialectical exploration of the adequacy of one or another of the alternative conceptions of emergence in theorizing about the explanation of human behavior. What we may suggest, then, heuristically, is this. Freud's theoretical orientation on the *Project* tends to favor the possibilities of the rejection of emergence in Feigl's sense as well as the admission of the sort of moderate concessions of emergence in Bunge's sense; but the development of his actual psychoanalytic work tends to be recalcitrant in the sense of the third sort of emergence. Gradually, the prospects of reductive explanation had come to seem very dim to Freud, regardless of his general adherence to materialism and his persistent hope that the main theme of the *Project* was correct.

Now, regarding the *Project* itself, the very first lines of the essay plainly state: "The intention is to furnish a psychology that shall be a natural science: that is, to represent psychical processes as quantitatively determinate states of specifiable material particles, thus making those processes perspicuous and free from contradiction. Two principal ideas are involved: (1) What distinguishes activity from rest is to be regarded as Q [some unspecified but quantifiable energy associated with the neurones], subject to the general laws of motion. (2) The neurones [the ultimate units of the nervous system] are to be taken as the material particles." Here, it is clear that Freud is attracted to something very much like Feigl's identity thesis expressed in physical$_2$ terms. He was tempted by a mechanical account (his own term) of psychical states. But almost at once he worries about its adequacy, and he begins to flirt with a kind of epiphenomenalism. Thus, he eventually notes difficulties "in introducing the phenomena of consciousness into the structure of quantitative psychology." And he adds: "According to an advanced mechanistic theory, consciousness is a mere appendage to physiologico-psychical processes and its omission would make no alteration in the psychical passage [of eeents]. According to another theory, consciousness is the subjective side of all psychical events and is thus inseparable from the physiological mental process. The theory developed here lies between these two. Here, consciousness is the subjective side of one part of the physical processes in the nervous system, namely of the ω processes; and the omission of consciousness does not leave psychical events unaltered but involves the omission of the contribution from ω.[14] The editor suggests that, under the influence of Hughlings Jackson, Freud was originally attracted to the "appendage" theory; but the fact remains that the very distinction of a "subjective side" of psychical events is both an unexplained puzzle and a distinct challenge to Freud's notion of a science. He himself says very clearly that "our consciousness furnishes only *qualities,* whereas science recognizes only *quantities* . . . [and] whereas science has set about the task of tracing all the *qualities* of our sensations back to *external quantities,* it is to be expected from the structure of the nervous system that it consists of contrivances for transforming external *quantity* into quality."[15] It is not at all clear, however, what can be meant by the subjective side of ω processes. That there are distinct kinds of neurones—ϕ (permeable neurones, that is, those external stimuli reach), ψ (impermeable neurones, that is, those involving "the discharge of excitations of endogenous origin"), and ω (perceptual neurons, perhaps a specialized subset of the ψ neurones, except that the ω neurones must be permeable, must behave like the sensory systems, in order to "generate conscious sensations of qualities")[16]—that there are distinct kinds of neurones, does not really help (as Freud himself notices) because it is the emergent property of consciousness (the "subjective side" of the neurones) that needs to be accounted for. Freud's appreciation of this difficulty, we may suppose, leads him to consider the alternatives of epiphenomenalism and something very much akin to Bunge's non-reductive kind of emergent materialism: this seems to be the only reaoonable sense to assign to the remark in which Freud weighs the alternatives of conceding the negligible omission of consciousness in scientific explanation as opposed to insisting on the inseparability of consciousness from the function of ω neurones. At the very least, at this point, the denial of emergence, in Feigl's sense, is clearly repudiated.

Having noted this much, it is but a step to what is very nearly Freud's last reflection on the matter. In the *Outline of Psycho-Analysis* (1938), Freud stresses that our starting point in analyzing psychical processes "is provided by a fact without parallel, which defies all explanation or description—the fact of consciousness"; in the same context, he criticizes American behaviorism for supposing that "it is possible to construct a psychology which disregards this fundamental fact!"[17] In context, this can only mean that not only the adequacy of physical$_2$ explanations but also the adequacy of any theory that might accommodate emergence in the restricted sense of Bunge's compositional thesis had become more than doubtful. In effect, therefore, Freud's increasing occupation with the actual details of the psychodynamics of human persons, as well as the explanatory power of psychoanalysis, led him inevitably in the direction of the third sort of emergence. And this, in spite of the fact that Freud never attempted systematically to clarify the conceptual relationship between his metapsychology and the explanatory nature of psychoanalysis and the underlying neuronal system.

The point is an important one and easily obscured. There are, in effect, two distinct lacunae in Freud's *Project* as far as the assimilation of psychoanalysis is concerned. For one thing, there is no way in which variable phenomenological qualities can be accounted for. Freud himself senses this, and his ingenious speculations about a variety of psychodynamic processes—including memory, instinct, cathexis, pain, primary and secondary processes, ego, will, reality-testing, dreams, ideas, hallucination, affect, hysteria, thought and speech—are themselves *exclusively* characterized in functional terms, that is, in terms of functional relations among systems of neurones. This means that, *once* admitting a particular discriminable quality, functional operations upon such data are, Freud thinks, explicable in terms of some adjustment of the general schema of the *Project*; but the identification and reidentification of such data appear to fall outside the competence of the system. Secondly, the processes of the *Project,* though they are obviously designed to accommodate intentional phenomena (for example, hysteria), fail to permit us to trace intentional phenomena qua intentional and fail to enable us to identify such phenomena in terms of the quantifiable properties of the neuronal system itself. This dual limitation is adumbrated most succinctly by Freud's remark that "it is only by means of such complicated and far from perspicuous hypotheses [regarding the flow and binding of Q and the functional distinctions of ϕ, ψ and ω neurones] that I have hitherto succeeded in introducing the phenomena of consciousness into the structure of quantitative psychology. No attempt, of course, can be made to explain how it is that excitatory processes in the ω neurones bring consciousness along with them. It is only a question of establishing a coincidence between the characteristics of consciousness that are known to us and processes in the ω neurones which vary in parallel with them. And this is quite possible in some detail."[18] But, of course, even the required parallel is impossible to establish without some viable account of how to relate determinate phenomenological qualities and intentional content to the determinate and quantifiable processes among the neurones. Otherwise, the paraphrase of psychopathological conditions in the language of the *Project* is entirely metaphoric.

That it is metaphoric is, in a sense, the upshot of Freud's doubts about the adequacy of the *Project*'s explanatory model. It is also pretty clearly the conclusion forced on us by Freud's discussion of Emma's hysterical compulsion, in Part II of the *Project*—really the only sustained attempt to link the language of psychopathology and the language of the *Project*. First, Freud observes that "hysterial patients are subject to a *compulsion* which is exercised by *excessively intense* ideas. An idea will, for instance, emerge in consciousness with particular frequency without the passage [of events] justifying it; or the arousing of this idea will be accompanied by psychical consequences that are unintelligible."[19] He adds that "our analyses show that a hysterical compulsion is *resolved* immediately it is *explained* (made intelligible). Thus these two characteristics are in essence one."[20]

Now, in the context of Emma's hysteria and, in general, in theorizing about the qualitative and intentional distinctions of consciousness, Freud tries to speak *alternatively,* sometimes even *jointly,* in an idiom in accord with his metapsychology and in an idiom adjusted to the *Project.* Freud's discussion of pain shows this very clearly.[21] In the same spirit, Freud observes, introducing Emma's case: "We have seen that hysterical compulsion originates from a pecular kind of $Q\eta$ motion (symbol-formation), which is probably a *primary process,* since it can easily be demonstrated in dreams; [and we have seen] that the operative force of this process is defense on the part of the ego, which here, however, is performing more than its normal function."[22] It must be borne in mind that such notions as those of symbol-formation, primary process, dreams, ego-defenses, ego have already been characterized by Freud, in Part I of the *Project,* as functional complexes of aggregates of neurones: in short, in a way that should have provided for explanations more or less in accord with Bunge's emergentism if not in accord with Feigl's physical$_2$ theories. But when he considers Emma's compulsion of not being able to go into shops *alone,* he provides a quite pretty analysis in terms of the psychodynamic idiom.

Freud does two things with the case. First of all, he resolves the puzzle of the hysteria in terms associated both with ordinary purposive behavior and with his distinctive extension regarding unconscious processes: "We now understand Scene I (shop-assistants) if we take Scene II (shopkeeper) along with it [he says]. We only need an associative link between the two. [Emma] herself pointed out that it was provided by the *laughing*: the laughing of the shop-assistants had reminded her of the grin with which the shopkeeper had accompanied his assault. The course of events can now be reconstructed as follows. In the shop the two assistants are *laughing*; this laughing aroused (unconsciously) the memory of the shopkeeper. Indeed, the situation had yet another similarity [to the earlier one]: she was once again in a shop alone. Together with the shopkeeper she remembered his grabbing through her clothes; but since then she had reached puberty.

The memory aroused what it was certainly not able to at the time, a *sexual release,* which was transformed into anxiety. With this anxiety, she was afraid that the shop-assistants might repeat the assault, and she ran away."[23] Here, Freud attempts to provide an explanation in terms of the purposive life of Emma, in terms of the *reasons* she behaved as she did, given her perception of the situation and assuming that she had, through unconscious associations, reasons for certain beliefs, affect, and motives. But at this point, he also asserts that "it is quite certainly established that two kinds of ψ processes are mixed up together here." Freud then attempts (the second move) to diagram the dynamics of the entire case, marking in those that he takes to involve ψ processes and those, ω processes. But the account is essentially psychodynamic. The only connection between the two idioms lies simply in tagging every mention of consciousness as ω processes and every mention of unconscious associations as ψ processes. The metaphoric quality lies in the implicit identification of "excessively intense" ideas with a certain rise in the level of concentration of bound Q. In context, however, he actually tends to favor the *Project*'s idiom rather than the psychodynamic idiom, in spite of the fact that it is the latter alone that *explains*—in any plausible sense at all—what happened to Emma.

This brings us round to the essential importance and fascination of Freud's entire endeavor. The issue is complicated by a consideration that arises from having noted the inherent difficulty of reconciling the explanatory enterprise of psychoanalysis with the scientific model preferred in the *Project*. Because, in supposing that Freudian psychology is a genuine science, one is led to suppose that it conforms with the canons that obtain among the physical or biological sciences—roughly, with those that obtain under Feigl's or Bunge's conception of a science. But *if* there are domains that cannot be explained (or adequately explained) in terms of such sciences, if there are domains that exhibit our third sort of emergence, it does not follow (i) that the phenomena of such domains cannot be explained; or (ii) that such explanations cannot form the basis of a distinctive science. The only thing that follows is (iii), that the various sciences (including, now, psychoanalysis) may not fall under the program of the unity of science.

The issue is nicely illustrated by some relatively recent quarrels about the status of Freudian analysis. In a spirited attack on the "pseudo-science" that Freud introduced, Frank Cioffi observes: "Freud is well aware that when a putatively pathogenic factor is discovered to be widely distributed this casts doubt on the genuineness of its relation to the illness. Before his abandonment of the seduction theory [that is, the theory that his patients' illnesses were due to their having been seduced in infancy] he said: 'Of course if infantile sexual activity were an almost universal occurrence it would prove nothing to find it in every case.' His awareness of this may account for his increased tendency to treat the patient's infantile sexual vicissitudes as *grounds for* rather than *causes of* his neurotic illness, since the explanatory status of a motive is not called into doubt because of the absence of a general connection between it and the behavior it is invoked to explain."[24]

There are a number of possible misunderstandings that can arise here. First of all, it is not the case that explanation in the sciences presupposes that all nomological connections be expressible as nomic universals. Statistical explanation is clearly admissible; and it is entirely possible that statistical laws reflect not an epistemic limitation regarding our grasp of the laws of nature but a metaphysical feature of nature itself.[25] Secondly, causal explanations of an agent's actions may be construed in terms of his reasons and motives, even if it is the case that there is a difference between explanation by reasons and explanation by causes.[26] And thirdly, *if* persons and their characteristic behavior are emergent in the third sense of emergence sketched, then it may well be the case that a distinct kind of explanation, a distinct kind of causal explanation, may be invoked in giving an account of human behavior. More pointedly put, it may well be that psychoanalytic explanation is and must be, because of the distinction of human existence, a form of explanation appropriate to the third sort of emergence; *and* that the distinction of such explanation lies in the particular way causal factors are informed by intentional considerations borrowed from the model of explanation by reasons. If this is so, we see the tendentiousness of Cioffi's remark that Freud's work fails to meet the qualifications of a science "because of the absence of a *general* connection between [a motive] and the behavior it is invoked to explain."

Cioffi's charge appears to make Freud's comment contrary, because it presupposes that the very feature Freud would concede to be uninformative is precisely what would be favored by the standard sciences; also, in virtue of this, Cioffi jumps to the conclusion that Freud cannot offer a causal account at all—and cannot, therefore, have formulated a science. Nevertheless, Cioffi himself cites quite a large number of assertions of Freud's, regarding libido theory, that tend to show (contrary to the spirit of Cioffi's own conclusion) that Freud was quite prepared to admit what appear to be statistical laws at least (if not nomic universals) governing the emotional life of human beings. There is a very simple explanation by which to reconcile Freud's constant mention of statistical regularities (that Freud himself was inclined to take as approaching invariance) and his remark that mention of such regularities does not explain a particular neurotic illness. The fact is that Freud wished to explain neurosis in terms of the quite determinate features of the history of his patients: mention of the uniformities he believed he had detected do not as such explain a particular neurosis. In fact, a reading of the *Introductory Lectures on Psycho-Analysis* (1916-1917) from which Cioffi quotes, clearly demonstrates that Freud construed his psychodynamic generalizations as lawlike regularities formulable in the functional terms posited by the schema of the *Project* at the same time they could not be more than merely determinable regularities from the point of view of the psychoanalytic method.[27] But the admission positively *requires* that the psychoanaltyic method exhibit an explanatory form noticeably different from that which obtains in the physical and biological sciences. Of the citations he provides, Cioffi says: "These statements certainly look like hypotheses."[28] Indeed they are, from Freud's point of view; but it is a *non sequitur* to conclude that they should play a direct explanatory role in psychoanalysis.

It needs to be said that psychoanalysts are themselves quite often misled by the complexity of Freud's position. This seems to some extent to have been true enough of Heinz Hartmann's well-known discussion of psychoanalysis as a science[29]—so much so that it led Ernest Nagel, for instance, to conclude that "psychoanalytic theory is intended to be a theory of human behavior in the same sense of 'theory' that, for example, the molecular theory of gases is a set of assumptions which systematizes, explains, and predicts certain observational phenomena of gases"[30]—and therefore to criticize it as a scientific theory.

Now, Hartmann *does* mention certain aspects of Freud's method that are *not* readily subsumable under the canonical view of a science. He remarks, for instance, that "Often historical explanations are substituted for system; an attempt is made to clarify the function of propositions in their relation to others by tracing their place in the development of analysis."[31] The point is easily misunderstood. What Freud apparently believed he produced in producing his metapsychology was a general set of dynamic uniformities of psychic development and functioning that transcended the biographical idiosyncrasies of individual patients; hence, also, that transcended the historical idiosyncrasies of particular societies and cultures.

Freud has, of course, been substantially criticized within the psychoanalytic movement as well as without. Within, the tendency, briefly put, has been to reject "the conviction that psychoanalytic propositions must be formulated in terms of Freud's metapsychology or in terms compatible with it."[32] Without, it has become quite standard to suppose that Freud was himself so captured by the orientation of his own historical experience that he generalized indefensibly about the *legitimate* regularities of the small and homogeneous sample that he was able to examine.[33] Roughly, the metapsychological generalizations, based on his own clinical experience, Freud took to be invariant (or at least to approximate actual invariances) because he believed he had discovered the invariant stages of the structural development of human beings.[34] But these very generalizations are intended to rest on empirical materials and *those* materials are inseparably linked to the "psychoanalytic narrative."[35] Hence, on any fair account, Freud must be supposed to have boldly generalized about the lawlike regularities of human psychodynamics *from materials that were originally not studied in the manner of the canonical sciences.* That is, Freud theorized in the canonical spirit of the *Project,* but his explanations are formulated in terms of the biographically relevant and causally efficacious beliefs, desires, and affect of particular persons in particular circumstances.

Here, we must lay down a number of essential distinctions without full defense. First of all, it is entirely reasonable to treat the reasons and desires that an agent has as causes of his behavior. The usual objection concerns an alleged threat to human freedom, which, without argument, depends on the unconfirmed assumption that causal regularities invariably presuppose determinism in accord with nomic universals. Second, there is no incoherence in extending the idiom of personal motivation to include unconscious processes; hence, there is no incoherence in speaking of the causal efficacy of unconscious motivation. The anthropomorphism of Freud's metapsychology may not unreasonably be construed as a heuristic convenience for functional distinctions relative to the molar behavior and ideation of human agents. Thus seen, Nagel's observation that "the notions of unconscious psychic processes possessing causal efficacies—of unconscious, causally operative motives and wishes that are not somatic dispositions and activities— . . . are just nonsense to me"[36] may be fairly offset by noting that apparent ascriptions to subpersonal systems (id, ego, superego) are invariably associated, in the psychoanalytic context, with explanations of the reasoned behavior of personal agents involving both conscious and unconscious processes. Third, to explain behavior as rational, in terms of the reasons for which what was done was done, is not to employ as such a species of causal explanation, though it is not to preclude causal explanation.[37] Briefly, reasons for acting can be specified for acts only when suitably described, that is, only in intensional contexts; whereas causes can be specified for events (not thereby explained) however described or identified, that is, extensionally. Furthermore, the specification of reasons presupposes a model of purposive life linking beliefs, intentions, desires, and action; norms of rationality; possibly both a schema of rational behavior that may be determinately realized in different ways by different cultures, as well as a historically or culturally oriented model suited to particular societies. If these distinctions be granted, then the causal explanation of characteristic human behavior may be said to fall under "covering institutions" rather than "covering laws," in the precise sense: (a) that human beings are indoctrinated into (causally affected by) the determinate institutions of a particular culture; (b) that they learn thereby to favor and disfavor actions viewed under particular descriptions and to act for suitably congruent reasons; (c) that covering institutions are sufficiently flexible to accommodate the variety of individual histories that tend, nevertheless, to converge toward a putatively normal range of rational behavior; and (d) that the institutions are themselves open to empirical inspection and subject to causal forces that change and replace them with other similarly influential institutions.

If these distinctions be admitted, then it is quite impossible that psychoanalysis should proceed empirically in such a way as to satisfy the canonical requirements that Nagel and Cioffi insist upon. On the other hand, there is good reason to believe that the model proposed is just the one that obtains in the cultural and historical sciences.[38] The heterodox feature of explanation in cultural contexts is just that (i) causes are identified individually; (ii) causal regularities may be subsumed under institutions rather than laws; (iii) lawlike and institutional regularities are entirely compatible.

Now then, the peculiarity of Freud's use of this model is just that, though his psychoanalytic work proceeds by considering (as Hartmann suggests) the life-*history* of a particular patient, in order to shape a completely adequate account of his reasons and motivation for behaving as he does, Freud is somehow persuaded (quite possibly by the very canonical form of the *Project*) that the anecdotal or narrative account of the behavior of a patient instantiates a deeper, invariant developmental pattern that belongs to the human species. Consequently, the kind of generalizations that Cioffi cites must be viewed as the *determinable* regularities that an entire range of determinate psychoanalytic explanations (keyed to individual idiosyncrasies) exhibit. Hence, too, the kind of charge that Nagel advances could be vindicated only if it were shown that psychoanalysis did not or could not proceed in a way that may fairly be called scientific—though in accord with our third kind of explanatory model (which is distinctly not canonical). For what the explanation of human behavior calls for is the evidence that an agent was causally influenced to act as he did (that is, under a description favored by him, in some appropriate sense) for reasons of a certain sort (also rightly assigned to him). In effect, explanation here is the explanation of *reasoned be-*

havior; of why acts performed are describable as they are from the agent's point of view; and why, thus described, they are performed for the causally efficacious reasons (conscious or unconscious) that they are.

Obviously, the complexities of our third kind of explanation need to be separately explored. We may take notice here only of the questions that remain to be resolved. For one thing, the model would have to be shown not to presuppose Cartesian dualism. For a second, the relationship between human actions and physical events would have to be explained. For a third, the coherence of a form of discourse that combined intensional reference to reasons and extensional reference to causes would have to be demonstrated. For a fourth, the distinctively cultural emergence of persons and their characteristic behavior would have to be conceptually linked to the admitted regularities of physical and biological phenomena. And for a fifth, the alleged logical properties of this third kind of explanation would have to be sorted out in order to make explicit the precise sense in which relevant hypotheses could be confirmed and disconfirmed.[39] These are not insurmountable complexities. In our present context, we need only assure ourselves that the proposal is not incoherent, has substantial implications for the unity of science and, a fortiori, for certain standard criticisms of Freud in the name of the unity of science, and provides a reasonably perspicuous model for the actual process of psychoanalytic explanation.

The central consideration is this. The behavior and ideation that the psychoanalyst attempts to explain are identified in intensional terms in a double respect. Not only is it the case that, as is generally admitted, explanatory contexts behave intensionally; in addition, *what* is to be explained is the already intentionally construed causal linkage between action and reason for action, that is, behavior conformable to some model of rationality. But this means that the action to be explained can be identified only under a suitable description; once identified, it may of course be reidentified under alternative descriptions, even nonintentional ones; and thus identified and reidentified, it may be assigned a certain causal force and discourse about its assigned causes may be taken to behave extensionally. But in the psychoanalytic context, in fact in the context of culturally significant behavior in general, we are to suppose that the agent was moved to act as he did for the reasons he did because he construed the act under one description rather than another; in the analytic context, it is further supposed that there may be found an alternative intentional characterization of action and motive (of an unconscious sort), suitably congruent with conscious motivation by way of some displacement or associative processes or the like, in terms of which a valid explanation of the third sort may be rendered. In short, the *causal* factors themselves cannot be taken to be efficacious in the sense appropriate to the third sort of explanation except in terms of some intentional characterization properly assigned to the psychological state of the agent—in a sense, in terms of what is "perceived" by the agent, whether consciously or unconsciously. This is perhaps the clue to that well-known remark of Freud's "*Psychic reality* is a special form of existence which must not be confounded with *material reality*,"[40] which Nagel cites disparagingly. It may be recalled here that Freud actually studied with Franz Brentano—may be taken to have extended, at least implicitly, the range of the intentional favored by Brentano in the context of conscious behavior and ideation.

The focus of these remarks is simply that the question of *which action* an agent has performed for *what reasons* cannot even be formulated except in terms initially of the model of explanation by reasons, not in terms of the model of explanation by causes. Hence, to give a causal account of reasoned behavior is to select intentionally informed causes by trading on the model of reasons. But though there may be statistical regularities linking causes and actions caused, these cannot be interpreted in terms of standing laws of nature but only of prevailing institutions, practices, modes of life subject to historical replacement; or better, whatever statistical laws obtain, the explanation of particular patterns of behavior cannot be given exclusively or even initially in terms of such regularities—the idiosyncratic life-history of the agent is required. We have only to recall Freud's explanation of Emma's hysteria to see the fit between our model and his actual way of working (apart from the later refinement of his methods). Emma's behavior constitutes one determinate incident in a family of determinate instances that exhibit the general form of the seemingly lawlike regularities that Cioffi cites. But two things must be noticed. For one, these would-be lawlike regularities cannot be confirmed in the canonical way but can be confirmed only by collecting a large number of convergent cases like Emma's, the explanation of each of which is antecedently confirmed in another way altogether. The reason is that the original act to be explained cannot even be suitably identified except under some intentionally qualified description: the problem of reidentifying one and the same action and the problem of the relationship of intentional actions and physical events are familiar enough.[41] The point is that it is impossible to reidentify one and the same *action* except in terms of some institutionalized (not merely physical) criteria; and it is impossible to provide an algorithm for specifying those and only those physical events that either are or entail particular actions, actions of a particular kind. But this means, precisely, that there is no way in which the causal regularities *extracted* from the range of psychoanalytic practice or from some other culturally significant domain[42] could be empirically confirmed in the way in which Nagel and Cioffi suggest. So they are right in thinking that psychoanalytic explanation does not succeed in meeting the constraints of canonical science; but they are wrong in supposing that there is not some alternative formulation that could so succeed. The second thing to be noticed is the logical distinction of the explanatory model actually used. The analyst tries to formulate an intentional description *under which,* on the clinical evidence, the agent may be said to have performed the action given (unconsciously motivated) and then tries to formulate congruent reasons for so acting (again, unconscious) that, furthermore, (i) conform with some model of rationality or conditional rationality (acting reasonably, once given one's beliefs—as Theodore Mischel observes), possibly also with some special theory of human motivation; and (ii) suitably conform with the conscious reasons and conscious characterization of the act in question and the entire life-history of the agent.

Freud, of course, was entirely aware of the charge that "what is advantageous to our therapy is damaging to our researches,"[43] that is, that interpretations given analytically are simply effective suggestions. But he counters quite reasonably, insisting that "whatever in the doctor's conjectures is inaccurate drops out in the course of the analysis"; and he himself proposes what is at once a theoretical and a pragmatic criterion of reasonable accuracy, namely, that "We look upon successes that set in too soon as obstacles rather than as a help to the work of analysis; and we put an end to such successes by constantly resolving the transference on which they are based."

In this sense, then, Freud is inexorably led, through his clinical experience, to a distinctive kind of explanatory model that, however grounded in the expectations of the *Project* and the model of canonical science, ultimately undermines the adequacy of the latter in the domain of the "human studies"[44] and tellingly points to the deep difference between the physical and the cultural disciplines. This may well be the signal of his most profound achievement.

The subtle development of Freud's position, then, effectively illuminates a final theme, the prospect of moral concern, which hinges on a double distinction. Moral discourse is at once intensional and pointless without the assumption of human freedom. It is intensional in the sense that whatever is approved or disapproved of is relevantly judged only under an appropriate description. And it depends on freedom in the sense favored but incompletely analyzed by Hume, that is, as involving both "liberty of spontaneity" and "liberty of indifference."[45] Furthermore, intensional considerations arise only for a creature capable of freedom, and freedom cannot but be formulated in terms of alternatives weighed under particular descriptions. It is clear that the radically extensional idiom of the *Project* could not have been hospitable to ethical distinctions; and that the development of analysis, by exposing the palpable forms of the loss of "spontaneity," permits us to recover the possibilities of the "liberty of indifference." Freud himself wrote very little about freedom and even professed to be a determinist.[46] But his entire endeavor can only be understood in terms of some form of self-liberation. He himself speaks in just these terms (rare enough for him) in explaining how one must detach himself from the Oedipal situation.[47] Furthermore, though he professes that neither science nor psychoanalysis are occupied with judgments of value, he affirms, in the same context, that, for instance, "the ethical demands on which religion seeks to lay stress need . . . to be given another basis; for they are indispensable to human society and it is dangerous to link obedience to them with religious faith." They are in fact to be linked to a *Weltanschauung* "erected upon science" which favors the "mainly negative traits . . . submission to the truth and rejection of illusions."[48] Apart from Freud's own ethical orientation,[49] what we need to grasp is that only an explanatory model like that afforded by psychoanalysis itself could possibly provide the conceptual link between scientific explanation and ethical appraisal.

If explanation in the sciences exclusively invoked nomic universals, then freedom would be precluded wherever such explanation obtained; if the sciences were restricted to physical$_2$ determinism or conceded emergence only with respect to compositional complexes of elements of some more fundamental physical level, then human persons could not be accounted for; and if the emergence of linguistic and other cultural abilities were not subject to scientific investigation, we could not regard the explanation of human behavior—formulated in terms of intensional distinctions involving reasons and purposes—as part of the enterprise of science itself. We need not accept Freud's system. Empirically, there may be much to complain about. But the charm of the *Project* is that it leads us, as it led Freud, to discover the inherent limitations of certain sorts of systems applied in contexts in which culturally relevant behavior calls for explanation. In effect, what Freud's work makes crystal clear— apart from the fortunes of psychoanalysis—are the conceptual constraints to which *any* rich examination of human behavior would have to conform. In this sense, it illuminates the conditions regarding the foundations of ethics; for ethical discourse functions only in contexts in which persons, their capacity for freedom, for choice, for deliberate and reasoned action, are neither eliminated nor reduced. This much, then, is neutral as between competing ethical doctrines; it tells us no more than what considerations must obtain if the conceptual foundations on which ethical claims are to be made may be reconciled with the entire enterprise of science.

A very powerful consequence of these concessions—ambivalently opposed to Freud's own convictions—is that there can be no discovery of ethical norms natural to human persons. If persons are culturally emergent entities, rather in the sense in which works of art are culturally emergent,[50] then the "nature" of human persons—however much it may depend upon and incorporate biological and psychological regularities—is shaped by the historical development of human institutions. If we must speak of man's essential nature, then it is as close to the truth as any judgment that may be ventured to hold that man's distinction lies in his resilient capacity to subscribe to historically widely variable and transient doctrines, values, ideologies, practices. This need not lead to ethical anarchy, conventionalism, subjectivism, skepticism, or arbitrariness. But it does suggest the inescapability of what may be called a "robust relativism," that is, the avoidance of radical relativism on the one hand and of some exclusive and objectively correct and discoverable norms of conduct on the other.[51] The point roughly is this. Relativism is incoherent to the extent that it attempts to make relative meaning or intelligibility, truth, consistency or coherence, and reference.[52] But there is no need at all for a viable relativism to go to such extremes. This is not to deny the utility of such distinctions as of "truth-in-L," alternative formal canons, alternative conceptual systems, the testing of theories as a whole, and the like. But the intelligibility of a common world, including the mutual intelligibility of historically divergent communities, argues the impossibility of making everything relative. Correspondingly, the validity of divergent ethical systems not only must accommodate these most fundamental constraints on intelligibility, but the pertinence of ethical concerns themselves sets looser limits on admissibly variable doctrines.

In particular, it calls for the recognition of the characteristic prudential concerns of human beings—the preservation of life, the reduction of pain, the gratification of desires and wants, security, and the like.[53] These statistically prevalent concerns define the relevance of ethical issues: wherever one affects, appears to affect, or is disposed to affect the prudential interests of another, ethical justification is reasonably indicated. They therefore set as well the determin*able minima* of ethical relevance. The determin*ate* specification in an ethically serious sense of any of these concerns depends on a commitment to a coherent doctrine that is in touch with the detailed historical life of a community. Here, the nature of ethical dispute is at once pertinently focused, not radically relativistic, and thoroughly dialectical in form. To specify a determinate dispute is to engage the historically evolved and historically significant features of social life; hence, the force of competing views cannot be grasped apart from a sensitive understanding of changing institutions.[54]

Now, then, the shift from covering-law explanations to covering institutions accommodates precisely this historical and dialectical focus. Freud seems not to have emphasized historical change because he favored more fundamental biological and psychological regularities. If such regularities were confirmed, they would relevantly qualify the plausibility of competing systems of values: Freud, rather ambivalently as we have seen, views psychoanalysis as value-free and also as guarding the principal values of human existence. But his own method of analysis, impossible to reduce to the method required by the *Project,* is actually responsive to the details of personal history and the dialectical alternatives reasonably supported by his own cultural milieu. He strains for some structural sequence of stages of psychodynamic development; but he explains the episodes of human life in terms that depend on the intensionally sorted contingencies of individual careers. It is in this sense, then, that Freud's own intellectual development yields the explanatory paradigm that is at once the distinction of the "human studies" and the condition on which a scientific foundation for ethics must ultimately rest.

NOTES

1. Richard Wollheim, "Introduction," in Richard Wollheim, ed., *Freud. A Collection of Critical Essays* (Garden City: Doubleday-Anchor Books, 1974).
2. "Project for a Scientific Psychology" (1895), in *The Standard Edition of the Complete Psychological Works of Sigmund Freud,* trans. by James Strachey et al., vol. 1 (1886-1899) (London: The Hogarth Press and the Institute of Psycho-Analysis, 1966). All references to Freud are to the *Standard Edition* (*SE*), unless otherwise specified.
3. Karl Pribram, "The Neuropsychology of Sigmund Freud," in Arthur J. Bachrach, ed., *Experimental Foundations of Clinical*

Psychology (New York: Basic Books, 1962), p. 443; cited by Robert C. Solomon, "Freud's Neurological Theory of Mind," in Wollheim, *Freud* . . . .

4. Irving Thalberg, "Freud's Anatomies of the Self," in Wollheim, *Freud* . . ., p. 154. Cf. also, Peter Alexander, "Rational Behavior and Psychoanalytic Explanation," *Mind,* 71 (1962): 326-41; reprinted in Wollheim, *Freud* . . .; and Theodore Mischel, "Concerning Rational Behavior and Psychoanalytic Explanation," *Mind,* 74 (1965): 71-78; reprinted in Wollheim, *Freud* . . . .

5. Cf. Otto Neurath et al., eds., *International Encyclopedia of Unified Science,* vols. 1-2 (Chicago: University of Chicago Press, 1955); Carl G. Hempel, "Explanation in Science and in History," in R. G. Colodny, ed., *Frontiers of Science and Philosophy* (Pittsburgh: University of Pittsburgh Press, 1962).

6. Cf. P. E. Meehl and Wilfrid Sellars, "The Concept of Emergence," in H. Feigl and M. Scriven, eds., *Minnesota Studies in the Philosophy of Science,* vol. 1 (Minneapolis: University of Minnesota Press, 1956).

7. Herbert Feigl, *The "Mental" and the "Physical." The Essay and a Postscript* (Minneapolis: University of Minnesota Press, 1958, 1967).

8. Ibid., pp. 10, 11.

9. In an unpublished paper, "Emergence and the Mind," to appear in *Neuroscience.*

10. Cf. Daniel Dennett, *Content and Consciousness* (London: Routledge & Kegan Paul, 1969); also, Joseph Margolis, *Persons and Minds* (Boston Studies in the Philosophy of Science) (Dordrecht: D. Reidel, 1978).

11. Cf. Wilfrid Sellars, "Philosophy and the Scientific Image of Man," in *Science, Perception, and Reality* (London: Routledge & Kegan Paul, 1963); Joseph Margolis, "On the Ontology of Persons," *The New Scholasticism,* 50 (1976): 75-84; and Margolis, "Proposals Toward a Theory of Persons," in Raphael Stern and Louise Horowitz, eds., *Ethics, Science and Psychotherapy* (New York: Haven, forthcoming).

12. Cf. P. F. Strawson, *Individuals* (London: Methuen, 1959).

13. I have explored this in a systematic way in terms of the relationship of embodiment, in *Persons and Minds.*

14. *S.E.,* 1: 310.

15. Ibid., p. 309.

16. Ibid., pp. 302-305, 309-10.

17. *S.E.,* 23 (1940): 157; cited by the editor of the *Project.*

18. *Project,* p. 311.
19. Ibid., p. 347.
20. Ibid., p. 348.
21. Ibid., pp. 296, 321. There is some uncertainty about Freud's distinction between 'Q' and '$Q\eta$'. '$Q\eta$' seems to designate quantifiable energy specifically associated with the nervous system or, even more narrowly, psychical energy; but the nature of the general energy 'Q' is not specified.
22. Ibid., p. 353.
23. Ibid., pp. 353, 354.
24. Frank Cioffi, "Freud and the Idea of a Pseudo-Science," in Robert Borger and Frank Cioffi, eds., *Explanation in the Behavioural Sciences* (Cambridge: Cambridge University Press, 1970), p. 487.
25. Cf. Wesley Salmon, "Statistical Explanation," in Wesley Salmon, *Statistical Explanation and Statistical Relevance* (Pittsburgh: University of Pittsburgh Press, 1975). It is quite noticeable that a great many who think of scientific explanation suppose that necessarily causal relations fall under nomic universals. This, for example, is an essential part of the resistance to construing explanation in terms of reasons, or beliefs and wants, as causal explanations; cf., for instance, Anthony Kenny, *Will, Freedom and Power* (Oxford: Basil Blackwell, 1975).
26. Donald Davidson, "Reasons, Causes, and Actions," *Journal of Philosophy,* 60 (1963): 685-700; but also Joseph Margolis, "Puzzles about Explanation by Reasons and Explanation by Causes," *Journal of Philosophy,* 67 (1970): 187-95.
27. *S.E.,* 16: 389.
28. Cioffi, "Freud and the Idea of a Pseudo-Science," p. 475.
29. Heinz Hartmann, "Psychoanalysis as a Scientific Theory," in Sidney Hook, ed., *Psychoanalysis, Scientific Theory, and Philosophy* (New York: New York University Press, 1959).
30. Ernest Nagel, "Methodological Issues in Psychoanalytic Theory," in Hook, Ibid., p. 38.
31. Hartmann, "Psychoanalysis as a Scientific Theory," p. 5.
32. Roy Schafer, *A New Language for Psychoanalysis* (New Haven: Yale University Press, 1976), p. 3.
33. Cf. Christopher Caudwell, *Studies in a Dying Culture* (London: John Lane, 1947); Kate Millet, *Sexual Politics* (Garden City: Doubleday, 1969); and V. N. Volosinov, *Freudianism: A Marxist Critique,* trans. I. R. Titunik (New York: Academic Press, 1976).
34. Cf. *New Introductory Lectures on Psycho-Analysis* (1933), *S.E.,*

vol. 22, especially lecture 31; also, *The Ego and the Id* (1923), *S.E.,* vol. 19.

35. Cf. Michael Sherwood, *The Logic of Explanation in Psychoanalysis* (New York: Academic Press, 1969), chaps. 5-6.
36. Nagel, "Methodological Issues . . .," p. 47.
37. *Contra* Davidson, cf. Margolis, "Puzzles about Explanation by Reasons and Explanation by Causes."
38. This may well be a way of redeeming the distinction between *Naturwissenschaften* and *Geisteswissenschaften*; cf. H. P. Rickman, ed., *Dilthey. Selected Writings* (Cambridge: Cambridge University Press, 1976).
39. Cf. Benjamin B. Rubinstein, "On the Role of Classificatory Processes in Mental Functioning: Aspects of a Psychoanalytic Theoretical Model," in Leon Goldberger and Victor H. Rosen, eds., *Psychoanalysis and Contemporary Science,* vol. 3, 1974 (New York: International Universities Press, 1975).
40. From *The Interpretation of Dreams,* cited from *The Basic Writings of Sigmund Freud,* ed., A. A. Brill (New York: Modern Library, 1938, p. 548.
41. Cf. Charles Taylor, *The Explanation of Behavior* (London: Routledge & Kegan Paul, 1964); Alvin I. Goldman, *A Theory of Action* (Englewood Cliffs, N.J.: Prentice-Hall, 1970).
42. Cf. Erving Goffman, *The Presentation of Self in Everyday Life* (Garden City: Doubleday, 1957). Goffman, apparently, would not concede the point.
43. *S.E.,* 16: 452. Cf. also the "Wolf-man" case, *S.E.,* 17: 80, cited by the editor of the *Introductory Lectures on Psycho-Analysis.*
44. Rickman's phrase, *Dilthey. Selected Writings.*
45. David Hume, *A Treatise of Human Nature,* III, ii, ii; cf. also, Anthony Kenny, *Will, Freedom and Power* (London: Basil Blackwell, 1975).
46. *Psychopathology of Everyday Life* (1901), *S.E.,* vol. 6.
47. *S.E.,* 16: 336-337.
48. *S.E.,* 22: 168.
49. Cf. Philip Rieff. *Freud. The Mind of the Moralist* (New York: Viking, 1959).
50. My own version of this theory appears in *Persons and Minds.* It is, however, not that version but the admission of cultural emergence itself that is decisive. The comparison between persons and works of art appears in *Art and Philosophy* (Atlantic Highlands, N.J.: Humanities Press, forthcoming).

51. I have formulated the thesis for the context of aesthetics, in "Robust Relativism," *Journal of Aesthetics and Art Criticism,* 30 (1976): 37-46. Its application for other domains is reasonably clear. For example, I have sketched the analogue in medicine, in "The Concept of Disease," *Journal of Medicine and Philosophy,* 1 (1976): 238-255; and have shown the need for such an analogue in ontology, implicitly, in "The Axiom of Existence: Reductio ad Absurdum," *Southern Journal of Philosophy,* 15 (1977): 91-99.

52. Possibly the clearest example of a wholesale relativism more than verging on intellectual anarchy is afforded by Paul Feyerabend, *Against Method* (London: NLB, 1975).

53. A fuller account of these *minima* of ethical relevance is offered in Joseph Margolis, *Negativities. The Limits of Life* (Columbus, Ohio: Charles Merrill, 1975).

54. This orientation—broadly speaking, "Hegelian"—converges to some extent with the views of Alasdair MacIntyre, "Objectivity in Morality and Objectivity in Science" and Steven Marcus, "The Origins of Psychoanalysis Revisited: Reflections and Consequences"; both are included in this volume.

5

Attitudes toward Eugenics in Germany and Soviet Russia in the 1920s: An Examination of Science and Values*

Loren R. Graham

THEORETICAL DISCUSSIONS of the relationship between science and values usually lead to the conclusion that, in a strict sense, science is value-free. If one confines one's attention to the intellectual content of scientific theory, and thereby excludes both the impact of technology on values and the influence of scientists as a political and social group, a persuasive case can be made that science is, indeed, neutral. There is no logical bridge between "is" and "ought."

It is obvious, however, that this form of exclusivist analysis of science overlooks some of the most historically important aspects of the relationship of science to modern culture. Explicitly, these examinations exclude links between science and technology as well as the connection between societal values and conditions for scientific research. Implicitly, these discussions exclude those arguments linking science to social value that are the most common in our imperfect world: the incomplete, semiscientific, possibly fallacious claims that often are advanced in the name of science, and the evaluations of these claims that every citizen has to make in order to decide about the relevance of science for social and political life.

In this essay I would like to examine and compare two separate but chronologically simultaneous episodes in the history of human genetics that involve all of these "second-order" links between science and values: the development in the 1920s of eugenic movements in Germany and Soviet Russia. Such a comparison produces unusual insights into the connections between science and political values.

The mention of the topic of human genetics in Germany and Russia in the early twentieth century will automatically bring to mind the issues of national socialist racial theories in Germany and Lysenko's theories of heredity in Soviet Russia. In the period which I am examining here, however, national socialism had not yet come to power in Germany and scientists of greatly differing political views discussed human genetics, including Marxists and socialists of various sorts, Catholics, liberals, and conservatives. In Soviet Russia, in the same years, no one had yet heard of Lysenko, and the spectrum of debate about human genetics was surprisingly broad. In Russia at this time there were Marxists who also were eugenists and saw no contradiction in their respective positions, and there were anti-Marxists who were Lamarckians. In both countries, then—Germany and Russia—I will be examining events in the periods before single-minded political concerns overruled scientific elements in the discussions about human heredity and its social implications.

How did it occur that this early complex period, in which so much uncertainty existed about the political value implications of theories of human heredity, was replaced in both countries by a later period in which Mendelian eugenic theories were usually linked to conservative views of society and Lamarckian theories were usually linked to left-wing socialist views of society? Was this process entirely a social and political phenomenon, essentially distinct from the scientific theories being discussed, or was there something intellectually inherent in each of the competing theories of heredity which supports a particular political ideology? If the answer to the latter question is affirmative, what do these inherent factors say about the allegedly value-free nature of science?

In both Weimar Germany and Soviet Russia, the peak of the debates over the social and political implications of theories of human genetics occurred in the mid-twenties, and reference was made in each case to ongoing debate in the other society. In both countries, moreover, Darwinism and social Darwinism had been familiar topics of discussion for decades, and the arguments of the twenties were set against this older background.

An example of the degree to which early German eugenists at first separated themselves from particular political viewpoints can be seen in the initial writings of Alfred Ploetz (1860-1940), a founder of the leading German eugenics journal and a central figure in the German eugenics movement of the twenties. As a young man Ploetz's interest in the genetic future of man was stronger than his commitment to any particular social or economic order, and he criticized both capitalism and socialism because of the contradictions which he saw between them and the ideals of eugenics. Ploetz, therefore, was no businessman or academic scholar defending economic competition on the basis of Darwinism, a model long familiar to us.[1] Ploetz was concerned to find a third way, a eugenic society that would put the good of the future above the comforts of the present.[2] Ploetz believed that socialism presented the prospect of a humanitarian society, and he praised the alleviation of economic suffering which he thought it could bring, but he saw it as thoroughly in contradiction with the principles of eugenics. If the "weak" were protected through social legislation guaranteeing employment, health, and life support, then they would propagate their kind in increasing numbers. On the other hand, Ploetz had little better to say for capitalism. Capitalism caused great human suffering—he cited Marx, Kautsky, Bebel and other socialists to that effect—and, furthermore, by giving advantages to the hereditary wealthy in the struggle for existence it hindered free natural selection.

From this analysis, one might guess that the most appealing form of society to Ploetz would be a somewhat softened capitalism in which individual competition still played the major role in determining success in any one generation. Ploetz, however, was both too complex and too humane to propose such a society. He seemed to favor all aspects of socialism except its alleged genetic consequences. The specter of open competition within each generation was offensive to him. Therefore, he struggled to find a way of combining the humanitarian and egalitarian ideals of socialism with the eugenic goal of a constantly improving biological base for society.

It might seem that in calling for a humane, eugenic, socialist society Ploetz was attempting to square a circle. However, he believed that he had found a way to reconcile his contradictory demands; he proposed that all people be protected by welfare legislation from the raw struggle for survival inherent in pure capitalism, as the socialists demanded, but at the same time he would ensure that each generation consisted of genetically healthy individuals by having each married couple select the "best" of their own germ cells for fertilization. Ploetz noted that the germ cells, or gametes, of each couple vary greatly in their genetic constitution; the important thing, from the standpoint of eugenics, was that the children be at least genetically equal to their parents. By "shoving selection back" to the prefertilization stage he hoped to achieve eugenic goals without tampering with individual human rights or interfering with natural, legal parenthood. As he wrote:

> The more we can prevent the production of inferior variations, the less we need the struggle for existence to eliminate them. We would not need it at all if in each generation we were able to ensure that the totality of variations was qualitatively somewhat superior, on the average, to that of the parents.[3]

It is easy to say that Ploetz's suggestion was naive and entirely impractical from the standpoint of the biological science and technology of his time. No one in the early years of this century had the faintest idea how to give an overall evaluation of the genetic quality of different gametes (nor can it be done now).

Nonetheless, Ploetz was making an important point which is more interesting to us now than it was when he wrote. The parents who decide today to abort a fetus found to be mongoloid through amniocentesis are following Ploetz's policy of "shoving selection back" to the prebirth stage in the name of humane principles, although not to the gametes themselves. And a few recent efforts have been made even to extend selection back to the gametes, as Ploetz suggested, not only for the determination of sex, but also for other specific genetic reasons. Controversial as some of these proposals are, we must grant that Ploetz was an early person to see that value consequences of applying the science of human heredity depend more on *when* and *how* it is applied than on the question of what is inherent in the science itself.

In the middle twenties, academic geneticists, anthropologists, and eugenists in Germany disagreed sharply over terminological issues concerning human heredity and race. Although these disputes may seem somewhat scholastic at first glance, they formed the essential theoretical background for the subsequent degradation of the science of human heredity in Germany. What occurred in this period was the gradual crystallization of political value links to specific biological interpretations.

The word eugenics is today usually considered a pejorative term and is even occasionally erroneously equated with national socialist doctrines. It is worthwhile noticing, therefore, that in Weimar Germany "eugenics" was considered by the protonational socialist publications and organizations as a kind of leftist deviation. The race hygienists followed the eugenics movements in other countries carefully, and pointed out frequently that the leaders of the movements there often failed to understand the racial significance of eugenics. Whether one used the term *"Rassenhygiene"* or *"Eugenik"* became in the late twenties a kind of political flag, with often the more right-wing members of the movement favoring the first term, the more left the latter.

The position of the German socialists and communists toward eugenics in the twenties was complex and interesting. Contrary to what we might assume, in the early and mid-twenties the leftist press did not define the debate as one over nature and nurture in which the nature side was inherently conservative from a political standpoint. The leftists noted that many of the eugenists opposed the traditionally conservative institutions of society—monarchies and the nobility—on the grounds that they were regressive from a genetic standpoint, with no correlation between influence and natural ability. And most Marxists prided themselves on their scientific outlook, with a knowledge of genetics a desirable goal. Furthermore, they were definitely in favor of reforming society, as were the eugenists. Therefore, many socialists and communists were supporters of eugenics.

The principle of responsibility to society as a whole lay at the basis of many of the German eugenic and early race-hygiene programs. The democratic socialist writers of the twenties often applauded this principle even when it was advanced by racist authors. Writing in 1923 in one of the oldest socialist journals, George Chaym observed, "Socialism certainly does not take a negative position toward race hygiene (*Rassenhygiene*), insofar as race hygiene concerns theoretical and practical measures for the improvement of the race or avoiding its debasement." Chaym continued that, correctly understood, race hygiene "belongs unconditionally in socialism's work program."[4]

If we examine the most prestigious journal of German social democracy in the twenties, *Die Gesellschaft,* we will find there many discussions of race hygiene and eugenics, and up to the end of the Weimar period the journal continued to support eugenics when it was divorced from extremist and racist interpretations.[5] Even the word *"Rassenhygiene"* continued to be used by many leftist writers.

The first major criticism of the race-hygiene movement to appear in *Die Gesellschaft* was an article in February, 1927, by Hugo Iltis, a biographer of Gregor Mendel.[6] The article revealed a great deal about socialist attitudes toward eugenics. Iltis was absolutely direct in his criticism of the majority of the German race hygienists. He said they were subverting science to politics, and he noted with regret that a number of prominent academic geneticists were among them. He criticized the typological description of races which was displacing the populational view and which attributed mental and ethical qualities to individual races. Iltis also saw clearly the dimensions of the threat; indeed, he said that "Race delusion was on the point of conquering German science." And he predicted that "The time will come when the race hatred and race obscurantism of our day will be what witchcraft and cannibalism are for us now: the sad vestiges of depraved barbarism."[7]

Surprisingly, not only did Iltis continue to support eugenics, but he even accepted the term *"Rassenhygiene."* He was convinced that the main problem with the field was the drawing of premature and inhumane conclusions on the basis of fragmentary data. In the future, he maintained, after "decades and perhaps centuries of hard work" we will create "the foundations of a genetic race hygiene," one that will serve the whole human race.[8]

Iltis also saw in Lamarckism a way of softening the hard facts of human genetics, and in that way he helped forge the links between leftist politics and Lamarckism that were growing in the twenties. He believed that only through the inheritance of acquired characteristics would it be possible to show that everything the socialists were working for—better education, better living conditions—would have a beneficial genetic effect.

Not all left-wing German eugenists supported Lamarckism, but the conservative race hygienists were delighted when they could hang the label of "Lamarckism" on their opponents; critics like Iltis became easy game for them. Fritz Lenz, one of the leaders of the German race hygienists, a person who would applaud Hitler even before he came to power, retaliated for the right-wing movement by dismissing the most prominent socialist and religious objectors to race hygiene as hopeless Lamarckians trying to dodge the facts of modern biology by creating the fiction of the inheritance of acquired characteristics.[9] Lenz was an academically qualified geneticist, and on the question of Lamarckism he exploited his advantage to the hilt. He pointed out that Paul Kammerer, a dedicated socialist and admirer of the Soviet Union, was trying to prove the inheritance of acquired characteristics with his experiments on the midwife toad. The experiments were exposed as fraudulent, after which Kammerer committed suicide. In the Soviet Union, Kammerer was made a hero, the subject of a popular film.

Lenz described all of this in his journal *Archive for Race and Social Biology,* and then turned the tables on his left-wing critics.[10] They accuse us, he said, of inserting our own values into biology, when actually they are the ones guilty of maintaining that biology contains humane principles, those of Lamarckism. *We* know, on the contrary, that science is value-free, he said.[11] We must follow the facts of human heredity wherever they take us, and those facts tell us that man will genetically degenerate unless the strong and the fit are given advantages in propagation.

The inheritance of acquired characteristics is a myth, he pronounced, and Mendelian laws alone govern human heredity.

Perhaps the most outspoken attack on the German race hygiene and eugenics movement to appear in the entire Weimar period was an article in 1928 in the communist journal *Under the Banner of Marxism*.[12] The author, Max Levien, presented a classical Marxist critique, simplistic but trenchant, and he avoided the Lamarckian trap. The central thesis of his long polemic was the view that the main current of German race hygiene was a bourgeois science which served the ruling class of German capitalists. The leading race hygienists preached a false theory of genetic degeneration because by portraying such a threat they hoped to justify imperial expansions against "inferior races and nations," particularly the eastern Slavs. Both of these goals were parts of their program of stemming the revolution already visible in the East.

Like the democratic socialists already discussed, the communist Levien believed that eugenics was a progressive, useful science. Once the revolution was victorious, eugenics could serve the proletariat as it had once served the capitalists. Levien said that German Marxists must not let the race hygenists do damage to "serious scientific efforts to create a people's eugenics (*Volkseugenik*) which considers the whole population of the earth."[13] Once race hatred has been deprived of its necessary base in the capitalist order, Levien asserted, man "will become a creative shaper of his own species; this advancement will bring about an upward surge in mankind's intellectual achievements and the power of science will thus enhance the rational, practical use of the laws of genetics for the development of an undreamt-of higher kind of man."[14] Levien's communist eugenics would be a part of an overall *Volkshygiene* program that would include both genetic and environmental improvement measures.

The relevance of Levien's argument for the Soviet Union is obvious. There the successful revolution he dreamed about had already occurred. The journal in which he wrote was the Western communist theoretical publication most widely circulated among Marxist intellectuals in the Soviet Union. Indeed, many of the articles were written by Soviet citizens. Would the Soviet Marxists, freed of the capitalist order, follow Levien's suggestion of developing a socialist eugenics? This will be one of the topics of the second part of this essay.

Observers of early Soviet history are frequently surprised to learn that Soviet Russia in the 1920s possessed a strong eugenics movement. One might have expected revolutionary Russia, which prided itself on its opposition to capitalist culture and aristocratic privilege, to have stood aside from the movement for "race betterment" which swept the world in those years, leading to the establishment of eugenics societies in several dozen different countries. To jump to this conclusion, however, is to carry back into the third decade of this century ideas both about eugenics and Soviet views of man which took clear form only in later years. Once again, as in Germany, we have the gradual formation of value links to scientific ideas, only in a very different political setting.[15]

In the first decades of the twentieth century, biologists in Russia had formed a center of outstanding genetics research. Around the figure of S. S. Chetverikov a school of population genetics was established which even yet is not fully appreciated by historians of biology, largely because it disappeared in later years when Lysenko took over Soviet genetics.[16] In addition to Chetverikov, outstanding early Soviet biologists included N. K. Kol'tsov, A. S. Serebrovskii, N. I. Vavilov, and Iu. A. Filipchenko. Several of these distinguished scientists were also heavily involved in the eugenics movement.[17] One of the outstanding younger Russian biologists would later emigrate to the United States, where he would become a world-famous scientist. He was Theodosius Dobzhansky, a scholar who began his career in Soviet Russia with a position in the Bureau of Eugenics of the Soviet Academy of Sciences.

The Russian eugenics movement was limited almost entirely to the years 1921 to 1930, and this period can be divided into two phases, with the division around the year 1925.[18] In the first phase, Russian eugenics developed along lines quite similar to the movement in a number of other countries. In the second phase, after 1925, Soviet eugenists made an effort to create a unique socialist eugenics of their own, an effort which met increasing opposition. Lysenko and the form of Stalinist genetics which later became notorious throughout the world were not involved in either of these phases, coming only in the late thirties and forties.

The two most important organizations in the Soviet eugenics movement were the Russian Eugenics Society and the Bureau of Eugenics of the Academy of Sciences, both created in 1921.[19] Each of them published a journal, the first entitled the *Russian Eugenics Journal,* the second (in its initial form) the *News of the Bureau of Eugenics.*

The historian who today leafs through the pages of the *Russian Eugenics Journal,* knowing well the class antagonisms and radical currents still waiting to be expressed in Soviet society, finds the naïveté and blindness to political complications of the early leaders of the movement quite striking. One of the early concerns of the authors in the journal in these years immediately after the Revolution was the genealogies of outstanding and aristocratic Russian families; investigations, complete with family tables, were made of princely families of exemplary achievements, as well as of all the members of the Academy of Sciences in the previous century. Several writers expressed dismay about the dysgenic effects of the Russian Revolution. The emigration of the nobility and of other upper-class families as a result of the Revolution was seen as a serious loss to the genetic reserves of Russia, requiring eugenic correction.

Several of the early Soviet eugenists showed that they were aware that their analysis and program were potentially inflammatory from the political standpoint. In the early spring of 1925 the Russian Eugenics Society debated eugenics proposals that had been advanced in other countries. Several Soviet participants objected to the "coarseness" and ill-defined quality of several of these proposals, some of which included plans for compulsory sterilization.[20]

The real significance of the Russian eugenics movement was not its relationship to Western eugenic proposals, but its place in the debates over the nature of man that were beginning to take place in Soviet Russia. How was the eugenics movement perceived by Soviet intellectuals outside the community of eugenists itself? What was the relation of the doctrines of eugenics to the ideas of Russian socialism and communism?

The great debate that eventually arose over the issue was rather slow in developing. The Commissar of Public Health, Nikolai Semashko, had given his approval to the eugenics movement, and the Commissariat of Internal Affairs (a police organization) formally accepted the charter of the Russian Eugenics Society. The Russian Eugenics Society received a small state subsidy.[21] These first official acts of recognition were probably not too significant in themselves, since in the early years the concept of eugenics was so new to most people in Soviet Russia, the essential issues so unexplored, that a sophisticated understanding of the movement by bureaucrats was hardly possible. To the extent that eugenics was understood it was thought to be the science for the collective improvement of mankind, and that was an activity that the young Soviet government automatically found interesting.

In 1925 the debate that had begun to simmer in lecture halls and in local publications spilled out into major Soviet intellectual journals. One of the first comprehensive and critical reviews of the eugenics movement was Vasilii Slepkov's article "Human Heredity and Selection: On the Theoretical Premises of Eugenics," which appeared in April, 1925, in the major Bolshevik theoretical journal.[22] Slepkov's point was that the eugenists were emphasizing biological determinants of human behavior to the total neglect of socioeconomic determinants. This point of view, said Slepkov, ignored the principles of Marxism, which demonstrated that social conditions determine consciousness. Slepkov quoted Engels on the importance of labor activity in the evolution of primates; Plekhanov on the view that conservative thinkers had always relied on explaining human behavior on the basis of innate qualities in order to avoid social analysis leading to revolutionary conclusions; and Marx to the effect that "people are a product of conditions and education and, consequently, changing people are a product of changing circumstances and different education."[23]

Slepkov did not deny, however, that individuals differ genetically. But in order to explain those differences he resorted to a view that was to have a long, and eventually tragic, influence on Soviet biology: the inheritance of acquired characteristics. He believed that social influence on the basis of conditioned reflexes, as he saw exemplified by Pavlov's research, was the best way of explaining the influence of environment on human behavior. And he pointed out that Pavlov believed that conditioned reflexes acquired during the lifetime of an organism could become, in some cases, permanently hereditable.

Despite Slepkov's emphasis, for many more years the inheritance of acquired characteristics remained as only a rather remote candidate for a replacement of classical genetics. As we will see, the geneticists and eugenists still had very powerful arguments at their disposal. Even in terms of politics, it was in the mid-twenties still not assumed in the Soviet Union that Marxist socialism and eugenics were incompatible, nor that Marxism and Lamarckism were uniquely compatible.

Nonetheless, not only the eugenists with their arrogant programs for biological reforms of society, but also the more sober and scientific geneticists interested primarily in animal and plant heredity were beginning to meet stiff opposition from radical students in Soviet universities. One biologist with a political commitment to the new regime, B. M. Zavadovskii, wrote that each year when he gave lectures at the Sverdlov Communist University the radical students reacted in a hostile manner to his discussions of heredity, calling genetics "a bourgeois science" containing implications that were unacceptable to the proletariat. Voices in favor of Lamarckism are becoming "louder and stronger," he wrote in 1925, as is the belief that genetics contradicts Marxism and the social policy of the Communist Party.[24] "This point of view is receiving support in the psychology of the masses," he wrote, "whose first reaction to genetics is negative." Zavadovskii feared that in dispensing with the erroneous views of the eugenists, Marxists might "throw out the baby with the bath" and eliminate the science of genetics as well. Marxists, he observed, have been "frightened" by the conclusions of the bourgeois eugenists.[25]

By 1925, then, the first crisis in the debate between Mendelian or classical genetics and Lamarckism was coming to a head, and the eugenists were in the middle of it. The eugenists began to defend both genetics and eugenics simultaneously, and they considered the Lamarckians to be one of the most important groups of their opponents.

Several of the eugenists realized that the criticisms being advanced against them were so serious that unless some way could be found of reconciling their understanding of human heredity with Marxist aspirations not only was the eugenic movement in danger, but genetics as well. Iu. A. Filipchenko, one of the brightest of the eugenists, found an argument in favor of genetics that carried great weight, and for a time was thought by some observers to have outmaneuvered the radical critics of genetics.[26] Let us assume for a moment, said Filipchenko, that the Lamarckians are correct, and the Mendelians incorrect, even though scientific evidence at the moment all points in favor of the classical Mendelian theory. What would be the result for the proletariat, for lower classes in general, and for the cause of social revolution if acquired characteristics were inherited? Most people seem to believe, Filipchenko observed, that such a theory points to rapid social reform. The same people usually consider classical genetics, with its stable genotype, as inherently conservative in its social implications.

This view is actually superficial and false, said Filipchenko, because it assumes that only "good" environments have heritable effects, while a consistent interpretation of the inheritance of acquired characteristics would show that "bad" environments have effects also. Therefore, all socially or physically deprived groups, races, and classes of people, such as the proletariat and peasantry and the nonwhite races, would have inherited the debilitating effects of having lived for centuries under deprived conditions. Far from promising rapid social reform, the inheritance of acquired characteristics would mean that the upper classes are not only socially and economically advantaged, but genetically privileged as well, a result of centuries of living in a beneficial environment, and one could never hope that the proletariat in Soviet Russia would be capable of running the state. It would be lamed genetically by the inheritance of the effects of its poverty.

If the classical geneticists were correct, on the other hand, said Filipchenko, then that would mean that distributed throughout the genes of the lower classes were combinations of genes that would give the individuals possessing them all the possibilities for being great scientists, musicians, artists, or whatever they might wish to be, if only the exploitative conditions hindering them up to this time were eliminated.

The logic of Filipchenko's arguments seems to have caused a hesitation in the criticism being advanced against the geneticists. Several radical journals ran articles maintaining that it was really the inheritance of acquired characteristics, not Mendelian genetics, that was counter-revolutionary. One author in such a publication stated that the international bourgeoisie constantly renewed efforts to establish the inheritance of acquired characteristics in order to show its own genetic superiority, but the proletariat was learning that science spoke against them. Another author, writing in *The Red Journal for All People,* said that every social reformer must read Filipchenko's argument in order to be armed for the political struggle.[27]

Filipchenko's argument was not persuasive enough, however, to stem permanently the popular belief that the inheritance of acquired characteristics was more congenial to the idea of creating a new society and a new culture than Mendelian genetics. And it should be noted that even those relatively few scientifically educated Soviet participants in the debate who listened to Filipchenko's view were not all convinced, some for fairly good reasons. After all, if Filipchenko *was* correct about the debilitating effects of an assumed inheritance of acquired characteristics on the lower classes, a great deal would depend on how long it would take to erase those effects—many generations or one or two. And, furthermore, the most that Filipchenko could promise on the basis of his classical genetics was that *some* members of the proletariat could excel because of their fortuitous possession of the right genes. The eventual result would be a class society based on innate, unchangeable ability, a meritocracy. The inheritance of acquired characteristics held a more radical, democratic prospect: *all* members of the lower classes and their progeny might advance equally because of their social environment, and, given sufficient time, there was no reason for the former lower class to be at any genetic disadvantage to the former upper class.

The overall result of these debates of the twenties was essentially a draw on the issue of whether the inheritance of acquired characteristics was theoretically more consonant with Marxism than classical genetics, although on a popular level, the belief among lay people and students that genetics was a bourgeois science continued to be strong and even grew as social conflict increased at the end of the decade with the commencement of the five-year plans.

The obvious losers in the debate were the Russian eugenists who defined their field in terms of the Western eugenics movement. They were not only supporters of classical genetics, with all the controversy that field attracted in the Soviet Union, but they went far beyond these theoretical principles to an extrapolation of biology to society, and in these extrapolations they had included a host of assumptions about the future of society, the nature of races and classes, and the relative influence of nature vs. nurture that did, indeed, conflict with prevalent views among the politically active elements of Soviet society.

The eugenists recognized that they must either abandon their concerns, or change radically their activities in order to demonstrate that they actually had the interests of Soviet socialism at heart, not Western capitalism. The scientists in the Bureau of Eugenics at the Academy of Sciences chose the prudent path of abandoning the field. Between 1925 and 1928 they shifted from a concern with human heredity and eugenics to a concern with the genetics of plants and animals.

The Russian Eugenics Society chose a more heroic and foolhardy path. The leaders of the journal decided to change their emphases, to show how eugenics could be fitted to the purposes of social revolution and Marxism. Turning away from their genealogical studies of the nobility, they began making studies of the reformers and revolutionaries of Russian history, beginning with the Decembrists of 1825, but continuing right up to contemporary Communist party leaders, whom they noted were not reproducing at an adequate rate.[28] Kol'tsov did a study of the genetic sources of the talented young proletarians being promoted up through the ranks of the universities and institutions of the Soviet Union.[29] Other eugenists began developing extended justifications of a unification of the goals of Soviet socialism and eugenics. But the whole effort became increasingly artificial and strained.[30] The eugenists, with their histories of connections with the international movement, including German race hygiene with its ever-clearer links to national socialism, could never justify themselves by trying to be more radical than their critics, or even equally radical, much as they were now trying. A scholar such as Filipchenko could win individual debates through the strength of his arguments about the social implications of an assumed inheritance of acquired characteristics, but by style and background he was always a middle-class *intelligent* to his critics.

By 1930 the eugenics movement in the Soviet Union was finished. These were years in which political controls in the Soviet Union were imposed in many scholarly institutions, and collectivization was violently enforced in the countryside. The Russian Eugenics Society was closed, and its publications suspended. By 1931, when the *Large Soviet Encyclopedia* published the volume of its first edition with an article on "Eugenics," the field was simply condemned as a "bourgeois doctrine."[31] By this moment in Soviet history the logical possibility of linking the Marxist desire to transform man with the eugenic desire to improve him had irretrievably disappeared.

By the early thirties the process of the gradual crystallization of value links to conflicting concepts of heredity—links that were at first by no means clear or inevitable—was far advanced in the two countries being studied here, Germany and Russia. Eventually the two societies went opposite directions in their interpretations of the nature-nurture controversy.

These two chronologically simultaneous episodes with contrasting results are probably as close to actual historical "test cases" on a large order of the question of whether different values are inherent in different theories of heredity as our very confused and inexact world is going to present to us. What do these examples tell us about the relationship between science and sociopolitical values?

Standing in our position today, it may appear that there is a natural alliance between eugenics and conservative, even fascist, sentiments. That link was not logically preordained, however, and was not perceived in the early twenties by large numbers of radical critics of society. Marxists and socialists of many types gave their support to eugenics in the early years, as did liberals, progressives, and conservatives. In the early twenties eugenics found response in both Weimar Germany and Soviet Russia across a rather wide range of political sentiment.

Eugenics in these early years was a faddish doctrine that was often considered progressive, the latest application of science for the benefit of humankind. If it was sometimes supported by aristocratic devotees of genealogical tables or middle-class members of social clubs, it was also on occasion given support by committed socialists who believed that a cultivation of true talent, rather than mere economic privilege, would destroy class society as it was known until that time. Both Weimar Germany and Soviet Russia were revolutionary states standing in the places of recent monarchies. Both in the twenties were going through eclectic periods when the full range of possible political implications of the latest scientific hypotheses had still not been formulated.

Was the passing of these early heterodox periods and the emergence of high degrees of consensus about the correlations between theories of heredity and political world views a phenomenon that touches science itself, or was it only an epiphenomenon of social and political turmoil? Do different theories of heredity intrinsically contain different value implications? The answers to these questions depend on what one considers the main rival theories to be. If one defines the contending theories in terms of their post-1933 vulgarized and absolutized forms (i.e., a national socialist theory of the "overall" values of different categories of humans, with Jews, gypsies, and certain types of "social misfits" genetically condemned to the bottom of the heap, and a Soviet alternative which opposes this view with a dogmatically environmental one), then it is obvious that the scientific "theories" in themselves contained by definition clear value components.

One can easily point out that these absolutized explanations, encased in official ideology and supported by the respective governments, can hardly be classified as "science." We will not use that fact, however, for a convenient exit from our dilemma about the relationship of science and values. We have purposely restricted ourselves in this study to the period before 1930, when professional geneticists and eugenists in both Germany and Russia usually defined the contrasting theories in much more academic terms. They often saw the difference between vulgarized extrapolations and core theories.

During the German Weimar and Soviet New Economic Policy (NEP) periods being considered here, the rival theories, stated in their scientific forms, did not explicitly contain value statements. The rival theories were those of Mendelian genetics and Lamarckism, viewed as scientific alternatives. Yet to many people they seemed to contain strikingly different value implications.

One of the several reasons for these apparent different value connotations stems from the state of genetics in the twenties as a pure and an applied science. Whether or not science *qua* science contains values, most people will agree that technology does have value impacts. Therefore, the significance for values of a particular science will vary as the associated technology for that science develops. One of the interesting aspects of the period of human genetics being studied here is that not only was human heredity understood in an exceedingly inexact fashion, allowing much room for speculation based on social and class motives, but there was no technology available for controlling the genetic constitution of organisms except by selection on the basis of phenotypes. Such selection was offensive to existing values when applied to people, and therefore humane individuals wished for an alternative even if they accepted the goal of more control over human genetics.

For this reason we take particular notice of the argument of the young Alfred Ploetz, still in his humanistic phase. He emphasized the fact that if the time ever came when selection could be shoved back before birth it might be possible to develop eugenic measures that were less offensive to traditional values than many of the proposals advanced in his day. We are now at a time when Ploetz's observation has a bit more meaning than it did when he wrote, at least for specific genetic defects. The differential impacts on values of various technologies thus serve as a further warning to us not to attribute values easily to scientific knowledge itself; the more important value-determinant is likely to be the technology derived from that knowledge.

The relatively few geneticists who did not fall prey to the political ideologies of the time in Germany and Russia were left with a very undramatic argument, one unable to carry their audiences in times of great social stress, of times when the active political elements were striving for answers to societal problems. The rigorous geneticists were saying, in effect: "Yes, the science of genetics is in principle applicable to man. However, genetics is so immature as a science that any effort to apply it to man now would have disastrous and unpredictable effects. Therefore, do not try to apply human genetics as a science, although continue to believe us when we tell you that science is ultimately the best hope for man."

In revolutionary moments popular audiences are not likely to find such an argument very persuasive. Both Germany and Russia were undergoing social and political upheaval, and the audiences turned toward more radical, although quite different, answers to the problem of heredity, answers that contained possibilities for application.

The German academic eugenists (as opposed to the race hygiene anthropologists), such as Fritz Lenz, were closer to science, as it existed at that time, when they attacked Lamarckists for their sentimental biology, than the Soviet Marxists who rejected genetics along with eugenics. The tragedy is that those scientists who—contrary to Lenz—remained loyal to genetics as a scientific theory while rejecting the growing inhumane and anti-rational extrapolations of German race hygiene were unable to find appreciative audiences in either Germany or Russia.

The reason for their failures was no doubt largely because of the social, political, and economic strains both societies were undergoing. To detail those strains would require a long digression into the social and political histories of Weimar Germany and Soviet Russia. But the existence of these strains was only a necessary, not sufficient condition for the emergence of radically different attitudes toward heredity. Science also was involved, a science of heredity that was so immature that it allowed room for wide speculation while offering few applications that were both reliable and ethical.

We are now approaching the core issue in the question of the relationship of science to values as revealed in our study of Germany and Russia in the twenties. We have already separated the question of political and social motivations from those of scientific inquiry to the maximum degree that is possible (it is never entirely possible) by focusing on more academic debates of the twenties, not the ideological exhortations of the thirties. We have separately considered the question of technology, trying to show where it had differential impacts on value that are distinguishable from those of scientific theory. Peeling away the layers of this problem, the central question emerges, one that relates to scientific theory alone: Was there something inherent in a hypothetical Lamarckian theory of heredity that made it a substantial buttress for egalitarian political values, and something within the Mendelian theory of heredity that lent support to elitist views of society?

Iurii Filipchenko tried to come to grips with this problem in 1926 when he maintained, with some success, that a hypothetical Lamarckian theory of heredity, followed to its conclusions, would mean the genetic impairment of those social classes whom left-wing social reformers wanted to help. The existence for many generations of the deprived classes in poor environmental conditions would supposedly have resulted in the acquisition of hereditary characteristics which would genetically impair them.

This argument was also used by the American geneticist and socialist, H. J. Muller, in his criticism of Lamarckism. (Muller and Filipchenko knew each other and the connection may not be coincidental). Yet we should admit that the argument by itself means very little. First of all, it is not completely clear that, in our hypothetical case of an assumed Lamarckism, deprived environments would lead uniquely to genetic damage; some people might maintain the reverse, seeing a sharp selection of the "fittest." Since Lamarckism is a doctrine without empirical support, there is no way of knowing how a hypothetical Lamarckism would fit with normal Darwinian selection. Furthermore, one would like to know how many generations would be required for a hypothetical Lamarckism to overcome the effects of earlier genetic impairment. Again, there is no way of knowing because we are not dealing with facts or testable hypotheses.

The assumptions that were carried by most Lamarckists into the debates in the Soviet Union of the twenties over human heredity were either (a) deprived environments do *not* result in impaired heredity, or (b) if deprived environments *do* result in impaired heredity, this impairment can be easily erased in a generation or two in the future through Lamarckian inheritance in improved conditions. *If* one makes these assumptions, then Lamarckian views truly fitted more comfortably with social reformism than Mendelian genetics, and this "science" contains real value implications. The Lamarckians could then promise the quick advent of a bright genetic future for all members of the proletariat (or any other group) while the Mendelians could promise such a future only to the fortunate few. This is the way the argument was usually seen by the thirties in the Soviet Union. Filipchenko lost the debate, the Lamarckians eventually won. But it should be clear that the chains of assumptions here took both of them into unreal worlds.

When we examine the relative value ingredients supposedly inherent in different theories of heredity, we should see that Filipchenko was not the only person to maintain that positive, humane values are inherent in theories of heredity that postulate a relatively stable genetic material, and that negative values lurk in the supposedly humane Lamarckian view of a malleable genetic base. Noam Chomsky has recently taken the view that man's relative immalleability is a protection for him against potential tyrants:[32]

> One can easily see why reformers and revolutionaries should become radical environmentalists, and there is no doubt that concepts of immutable human nature can be and have been employed to erect barriers against social change and to defend established privilege.
>
> But a deeper look will show that the concept of the "empty organism," plastic and unstructured, apart from being false, also serves naturally as the support for the most reactionary social doctrines. If people are, in fact, malleable and plastic beings with no essential psychological nature, then why should they not be controlled and coerced by those who claim authority, special knowledge, and a unique insight into what is best for those less enlightened?

This statement contains not only an observation on aspiring genetic manipulators of man, but also the concern of a more recent generation about the control of the behavior of man. From the standpoint of Chomsky's position, both the eugenic selectionists of Germany and the Soviet "makers of a new Soviet

man" represented threats of a fundamental sort to human freedom. Yet it is clear that Chomsky's observation, valuable though
it is, presents only one more insight, and is not an attempt to
give an ultimate analysis of whether putative theories about the
mutability of man, genetic or behavioral, contain inherently positive or negative value. He admits that the theories can work both
ways, with hypothetical immutability "employed to erect barriers
against social change and to defend established privilege" and
hypothetical mutability serving as "the support for the most
reactionary social doctrines." *Which* way the theories would work
in a given historical situation depends on the values of the
political and scientific authorities who disposed of the theories
and the associated technology.

It is my opinion that in our present state of knowledge of
human heredity the question of whether available theories have in
themselves positive or negative value content simply cannot be
answered on an abstract level, apart from reference to the existing sets of social forces and to the available technologies.
However, in a given historical situation, such as the ones we
have been studying here, rival scientific theories always exist
within the context of given sets of social and political circumstances, of competing political ideologies, and economic motives,
and of existing systems of technological capabilities. Within
those frameworks of external factors, rival scientific theories *do*
have differential value implications, but they derive their value
meaning much more from their relationships to these external
factors than from anything inherent in the science.

In terms of impact, then, we are not asking the right question
when we ask, "Do scientific theories contain in themselves social
or political values?" Instead, we should pay more attention to
what I called at the beginning of this paper the "second-order"
links between science and values, those which are contingent on
existing political and social situations, current technological capabilities, and the persuasiveness of current ideologies, flawed in
an intellectual sense as they may be.

If we look back at Germany and Russia in the twenties with
these overall extra-scientific frameworks in mind, it is clearly far
from accidental that the societies eventually ended up on the
particular opposite sides of the nature-nurture dispute with which

we now associate them. Arguments for radical egalitarianism *could* be made more easily and more persuasively in accordance with Lamarckian environmentalism, and arguments for hierarchies of social and racial values *could* be placed more comfortably within the Mendelian system. The more obvious points of debate leaned in these directions, while the more subtle ones were not heard at moments of social and political trial. In this global, approximate, and intellectually flawed sense, then, these rival scientific theories *did* contribute heavily to arguments about values. These value implications derived from the relationships of the theories to their social and economic milieus, but that is normally the situation with science.

Since I have maintained in this essay that the theories of human heredity current in Soviet Russia and Weimar Germany in the twenties contained in themselves much less inherent value component than is often considered to be the case, my viewpoint might easily be considered as a buttress for the argument that science is really value-free. My interpretation might even be used to attempt to free scientists from a sense of responsibility to society. My reply to this misunderstanding of my analysis would be to point out that the value-free interpretation of these theories of human heredity is persuasive only if the links of science and society are severed, and if science is viewed in abstract isolation from its setting. In actual fact, every scientific theory and every technological innovation *always* exists in a social and political setting, and the value impacts of these combinations can be massive.

Nuclear science is, seen in isolation, value-free. If, however, Nazi Germany had developed and employed atomic weapons in 1942, the configuration of our world might be very different today, and the effects on our values would probably be incalculable. Scientific descriptions of our universe—e.g., the Copernican or Ptolemaic alternatives—are, in the abstract, value-free, but the new and successful Copernican variant had a very large value impact when absorbed by a European civilization at a time when the older variant was firmly interwoven with religion and culture.

Scientific theories and technological innovations are thrust upon societies in which they may either buttress or counter existing value preferences. The powerful groups in society are usually more successful in turning science and technology to their advantage than the weak ones, so there is a natural tendency for science and technology to buttress the values of ruling classes or political groups. In both Germany and Russia by the thirties these links between ruling value systems and contrasting theories of heredity had been successfully forged, after some initial uncertainties, and the links then became powerful weapons of propaganda.

The responsibility of scientists for their creations is not less because of the fact that the value impacts of their work usually derive from a changing relationship between science and society, but greater. They must not only try to judge what *can* be done on the basis of their work but what, in all probability, *will* be done in view of the existing social forces.

As a hypothetical case, let me assume, somewhat grandiosely, that I am a scientist in Nazi Germany and that I have just discovered Tay-Sachs disease, a disorder with genetic causes that is more common among Jews than other groups of the population.[33] Whether I should publish my research and to whom I should send reprints (Hitler?) become moral acts, not because of values inherent in my scientific work, but because of the possible impact in that particular political setting of this purely scientific (not technological) finding.

When physicists speak of physical forces, they often speak of "strong" and "weak" forces. It might be useful for us to consider, for a moment, "strong" and "weak" value components in scientific theories. From the standpoint of a philosopher, the value components that I have found in scientific theories are weak, since they come not from the scientific theory itself, but from the relationship of the theories to society. Thus, Kenneth Schaffner commented on an earlier version of my essay:

> Professor Graham's view of the relations between science and values then seems to me to be an extremely *weak* one. Perhaps this is the one which he would in fact wish to defend. In my view, a much stronger thesis of the non-value-free character of science, including *basic* science, is defensible, but I do not have the time, nor would it be appropriate to develop my views here.

Since Professor Schaffner does not go on with his stronger

thesis, I can only speculate on what it might be. Certainly, if one can identify value components within science itself (and not in its relationship to society), this identification would result in a stronger case in philosophical terms (although we do not yet know whether the social significance of this value ingredient would be great or small). I suspect that such internal value components in some types of scientific theories *can* be identified. I am not, therefore, maintaining that science qua science is value-free. When, as others in this book have pointed out, scientists use terms such as "normal" and "abnormal" in physiology or psychology the importation of values seems likely. And doubtless there are other examples of diverse kinds in which value terms reside in the core of science itself.

Although I agree that the discovery of value components internal to science would be a stronger case in terms of philosophical considerations than the "contingent" or "relational" argument I have made here, I maintain that the value-impacts on society coming from science of the type I have been examining are far stronger in terms of the concerns of historians or sociologists. In other words, I question the social importance of the value components that may be truly internal to science, but I underline the importance of those that derive from the relationship of science to society.

Thus, I suspect that Professor Schaffner and I are using the terms "strong" and "weak" in two different ways. He is interested primarily, it seems, in strong arguments showing value ingredients in science, whatever the social significance of those components may be. I am interested in strong value impacts of science and technology on society. Therefore, in many cases, what he would call "strong" I would call "weak" and vice versa.

In my opinion, the really massive impacts of science and technology on social values are all of the type I have described here, i.e., the impacts that derive from the relationship of a particular scientific theory or technical innovation to a particular society in a particular historical setting. These impacts are the type that change history: they result in life and death, in vast changes in social mores, in changing attitudes toward the scientific enterprise itself. On the other hand, those values that may be truly inherent in science, built into its structure, are, in my opinion, of less significance, so much so that I would call them "weak."

One last word on the relevance of the episodes in Weimar Germany and Soviet Russia to the present day. . . . I have emphasized that the value contributions or inputs of the rival scientific theories depended much more on external social forces and available technologies than they did on anything inherent in the theories. We now live in a time when the social circumstances and available technology connected with the science of human genetics are very different from what they were in the twenties. We should not assume, therefore, that the value links of our time on the nature-nurture issue are identical to what they were in the twenties. To ignore the genetic basis of human beings at the present time, when we understand that basis much better and are in some instances able to affect it in reliable ways or alleviate great suffering, would be quite inhumane. Eugenics is now a word in disrepute, but the use of genetic knowledge to benefit mankind is a far more viable possibility now than ever before. On the other hand, some of the dangers of the twenties are still with us. We need to examine each issue relating to human genetics separately, considering the whole context of second-order value links, and decide the issues without a priori ideological commitments.

NOTES

* I would like to express my appreciation to the Rockefeller Foundation for a Humanities Fellowship for 1976-1977, during which time I did research on the relationship of science and sociopolitical values, including this article. I would also like to thank Harvard University for providing research facilities during the year, particularly the Program on Science and International Affairs, the History of Science Department, and the Russian Research Center. A longer and somewhat different version of this chapter appeared in December, 1977, in the *American Historical Review*.

1. There is a very large literature on Darwinism and social Darwinism in Germany. See, particularly, Hans-Günter Zmarzlik, "Der Sozialdarwinismus in Deutschland als Geschichtliches Problem," *Vierteljahrshefte für Zeitgeschichte* no. 11 (1963), 246-73; and the

same author's "Der Sozialdarwinismus in Deutschland—Ein Beispiel für den gesellschaftspolitischen Missbrauch naturwissenschaftlicher Erkenntnisse," in Günter Altner, ed., *Kreatur Mensch: Moderne Wissenschaft auf der Suche nach dem Humanum* (Munich: Moos, 1973), pp. 289-311. Also, see William M. Montgomery, "Germany," in Thomas Glick, ed., *The Comparative Reception of Darwinism* (Austin: The University of Texas Press, 1974), pp. 81-116; Gerhard Heberer and Franz Schwanitz, eds., *Hundert Jahre Evolutionsforschung: Das wissenschaftliche Vermächtnis Charles Darwins* (Stuttgart, 1960); and Hedwig Conrad-Martius, *Utopien der Menschenzüchtung: Der Sozialdarwinismus und seine Fölgen* (Munich: Kösel-Verlag, 1955). An interesting but somewhat simplified account is Daniel Gasman, *The Scientific Origins of National Socialism: Social Darwinism in Ernst Haeckel and the German Monist League* (New York: American Elsevier, Inc., 1971). Also, Niles R. Holt, "Monists & Nazis: a Question of Scientific Responsibility," *The Hastings Center Report*, 5, no. 2 (April, 1975): 37-43.

2. Alfred Ploetz, *Die Tüchtigkeit unserer Rasse und der Schutz der Schwachen: Ein Versuch über Rassenhygiene und ihr Verhältnis zu den humanen Idealen, besonders zum Socialismus,* (Berlin: S. Fischer, 1895).

3. Ibid., pp. 224-25.

4. *Sozialistische Monatshefte,* 10 (1923): 638.

5. In addition to the articles and reviews specifically discussed in the text, see the following reviews in *Die Gesellschaft*: R.F. Fuchs, review of A. Basler, *Einführung in die Rassen- und Gesellschafts-Physiologie,* 3, no. 3 (March, 1926): 283-86; Karl Kautsky, review of Alfred Grotjähn, *Die Hygiene der menschlichen Fortpflanzung: Versuch einer praktischen Eugenik,* 4, no. 4 (April, 1927): 381-82; M. Kantorowicz-Kroll, review of Robert Sommer, *Familienforschung, Vererbungs- und Rassenlehre,* 5, no. 7 (July, 1928): 92-4; Miron Kantorowicz, review of Ernst Neumann, *Individual-Rassen- und Volkshygiene,* 8, no. 9 (September, 1931): 288.

6. Hugo Iltis, "Rassenwissenschaft und Rassenwahn," *Die Gesellschaft,* 4, no. 2 (February, 1927): 97-114.

7. Ibid., pp. 108, 114.

8. Ibid., p. 113.

9. For examples of Lenz's attacks on Lamarckism see: review of Hermann Paull, *Wir und das kommende Geschlecht, Archiv für Rassen- und Gesellschaftsbiologie,* 15, no. 3 (1924): 330-32; review of Wilhelm Schmidt, *Rasse und Volk, Archiv für Rassen- und*

Gesellschaftsbiologie, 21, no. 1 (1928), 111-15; review of Friedrich Hertz, *Rasse und Kultur, Archiv für Rassen- und Gesellschaftsbiologie*, 18, no. 1 (1926): 109-14. An example of Lenz's praise of Hitler before national socialism came to power is the following 1931 statement: "Hitler ist der erste Politiker von wirklich grossem Einfluss, der die Rassenhygiene als eine zentrale Aufgabe aller Politik erkannt hat und der sich tatkräftig dafür einsetzen will." Fritz Lenz, "Die Stellung des Nationalsozialismus zur Rassenhygiene," *Archiv für Rassen- und Gesellschaftsbiologie*, 25, no. 3 (1931): 300-308.

10. F. Lenz, "Der Fall Kammerer und seine Umfilmung durch Lunatscharsky," *Archiv für Rassen- und Gesellschaftsbiologie,*21, no. 3 (1929): 311-18.

11. Lenz many times made the argument that science was value-free. For example, in 1921 he said, "Die Natur verlangt überhaupt nichts; die Naturwissenschaft kann nur zeigen, was geschieht, nicht was geschehen soll." *Archiv für Rassen- und Gesellschaftsbiologie*, 13, no. 1 (1921): 112.

12. Max Levien, "Stimmen aus dem teutschen Urwalde (Zwei neue Apostel des Rassenhasses), "*Unter dem Banner des Marxismus*, 2, no. 1/2 [4/5] (1928): pp. 150-95. The article is a criticism of the views of H.F.K. Günther and A. Basler.

13. Ibid., p. 155.

14. Ibid., p. 163.

15. For background on the reception of Darwinism in Russia, see James Allen Rogers, "Charles Darwin and Russian Scientists," *The Russian Review*, 19 (1960): 382. There was even an early advocate in Russia of improvement of humans through selection: V.M. Florinskii, *Usovershenstvovanie i vyrozhdenie chelovecheskogo roda* (St. Petersburg, 1866). For a discussion of this book by an early Soviet eugenist, see M.V. Volotskoi, "K istorii evgenicheskogo dvizheniia," *Russkii evgenicheskii zhurnal*, 2, no. 1 (1924): 50-55. Also see George Kline, "Darwinism and the Russian Orthodox Church," in Ernest J. Simmons, ed., *Continuity and Change in Russian and Soviet Thought* (Cambridge, Massachusetts: Harvard University Press, 1955), pp. 307-28; N. Danilevskii, *Darvinizm, kriticheskoe issledovanie* (St. Petersburg, 1889); K.A. Timiriazev, *Charlz Darvin i ego uchenie* 3rd ed. (Moscow, 1894); Alexander Vucinich, "Russia: Biological Sciences," and James Allen Rogers, "Russia: Social Sciences," in Thomas F. Glick, ed., *The Comparative Reception of Darwinism* (Austin: University of Texas Press, 1972), pp. 227-68.

16. See Mark B. Adams, "The Founding of Population Genetics: Con-

tributions of the Chetverikov School, 1924-1934," *Journal of the History of Biology,* 1, no. 1 (1968): 23-39; "Towards a Synthesis: Population Concepts in Russian Biological Thought, 1925-1935," *Journal of the History of Biology,* 3, no. 1 (1970): pp. 107-29.

17. N.K. Kol'tsov was president of the Russian Eugenics Society. Iu. A. Filipchenko was the director of the Bureau of Eugenics of the Academy of Sciences. A.S. Serebrovskii was a member of the permanent bureau of the Russian Eugenic Society, and a contributor to its journal. For his great hopes for the eugenic movement, see A.S. Serebrovskii, "O zadachakh i putiakh antropogenetiki," *Russkii evgenicheskii zhurnal,* 1, no. 2 (1923): 107-16, especially p. 112. Ironically, N.I. Vavilov, who ultimately suffered the most at the hands of Soviet authorities, apparently steered clear of eugenics. Vavilov became a foe of Lysenko in the late thirties and died in 1940 in Siberian exile.

18. A good bibliography of Russian eugenic literature is K. Gurvich, "Ukazatel' literatury po voprosam evgeniki, nasledstvennosti i selektsii i sopredel'nykh oblastei, opublikovannoi na russkom iazike do 1/I 1928 g.," *Russkii evgenicheskii zhurnal,* 6, nos. 2-3 (1928): 121-143.

19. For accounts of the founding and early activities of the Russian Eugenics Society, see "O deiatel'nosti Russkogo Evgenicheskogo Obshchestva za 1921 god," *Russkii evgenicheskii zhurnal,* 1, no. 1 (1922): 99-101 and similar descriptions in succeeding volumes of the same publication. For an interesting but somewhat one-sided recent Soviet interpretation of these early eugenic interests, see N.P. Dubinin, *Vechnoe dvizhenie* (Moscow: 1973).

20. "Obsuzhdenie Norvezhskoi evgenicheskoi programmy na zasedaniiakh Leningradskoi Otdeleniia R.E.O.," *Russkii evgenicheskii zhurnal,* 3, no. 2 (1925): 139-43. The particular eugenic program that was at the basis of this discussion was that of the Norwegian J.A. Mjöen. This discussion of the Russian Eugenics Society is very interesting, with separate consideration of "positive and negative race hygiene proposals," and identification by speakers of wide disagreements on issues such as racially mixed marriages and mandatory sterilization.

21. For approval by the commissariats of health and education, as well as for announcement of the subsidy, see "Iz otcheta o deiatel'nosti Russkogo Evgenicheskogo Obshchestva za 1923 g.," *Russkii evgenicheskii zhurnal,* 2, no. 1 (1924): 4. For approval of the society's charter by the Commissariat of Internal Affairs, see "Evgenicheskie zametki," *Russkii evgenicheskii zhurnal,* 2, no. 1 (1924): 58.

22. V. Slepkov, "Nasledstvennost' i otbor u cheloveka (Po povodu

teoreticheskikh predposylok evgeniki)," *Pod znamenem marksizma,* no. 10-11 (October-November, 1925), pp. 79-114.

23. Ibid., p. 113.

24. B.M. Zavadovskii, "Darvinizm i lamarkizm i problema nasledovaniia priobretennykh priznakov," *Pod znamenem marksizma,* no. 10-11 (October-November, 1925), pp. 79-114.

25. Ibid., p. 101.

26. See the discussion of the impact of Filipchenko's argument in "Spornye voprosy evgeniki," *Vestnik kommunisticheskoi akademii 20 (1927): especially 224-25.*

27. "Spornye voprosy evgeniki," *Vestnik kommunisticheskoi akademii* 20 (1927): 225.

28. Kol'tsov observed in a 1924 article, "If we calculate the average number of children in the family of each member of the Russian Communist Party that number will no doubt be far from what Gruber cites as necessary for a population group to preserve itself in the overall population. What would we say about a stock-breeder who every year castrated his most valuable producers, not permitting them to multiply? But in cultured society approximately the same thing is occurring before our eyes!" N.K. Kol'tsov, "Vliianie kul'tury na otbor v chelovechestve," *Russkii evgenicheskii zhurnal,* 2, no. 1 (1924): 15. The reference is to Max Gruber, *Ursache und Bekämpfung des Geburtenrückgange im deutschen Reiche* (Munich, 1914).

29. N.K. Kol'tsov, "Rodoslovnye nashikh vydvizhentsev," *Russkii evgenicheskii zhurnal,* 4, nos. 3-4 (1926): 103-43; for other examples of this kind of article, see N.P. Chulkov, "Genealogiia dekabristov Murav'evykh," *Russkii evgenicheskii zhurnal,* 5, no. 1 (1927): 3-20.

30. The difficulty of the position of the academic eugenists was illustrated by Serebrovskii when he called for a socialist eugenics and then observed: "Every class must create its own eugenics. However, this slogan . . . must in no way be understood in the manner of several of our comrades, especially in Moscow, who maintain that the whole base of Morganist-Mendelian theory is an invention of the Western bourgeoisie, and that the proletariat, creating its own eugenics, must base itself on Lamarckism." A.S. Serebrovskii, "Teoriia nasledstvennosti Morgana i Mendelia i marksisty," *Pod znamenem marksizma,* no. 3 (March, 1926), p. 113.

31. "Evgenika," *Bol'shaia Sovetskaia Entsiklopediia,* 23 (Moscow: 1931), cols. 812-9.

32. Noam Chomsky, *Reflections On Language,* (New York: Pantheon Books, 1975), p. 132.
33. I am grateful to Professor Noretta Koertge of Indiana University for suggesting this example.

Commentary

Science and Values—
Internal and External
Relations: Response to
Loren Graham

Kenneth F. Schaffner

I FOUND PROFESSOR GRAHAM'S PAPER "Eugenics and Human Heredity in Weimar Germany and Soviet Russia in the 1920s: An Examination of Science and Values" to be a beautifully balanced and intellectually stimulating essay. I particularly liked the way it related·the science of eugenics to the different cultural histories in the two countries under consideration. However, in spite of my admiration for the historical scholarship, I found myself disagreeing with some of the distinctions he made, and with some of the stress that he placed on what he terms "second-order" interactions between science and values. My comments will focus on these distinctions and stresses, and attempt to conceptualize the relation between science and the foundation of ethics in a somewhat different and I think broader manner.

My disagreements with Professor Graham's philosophical construal of the relation between science and values grows out of the following kinds of comments that he makes in his essay.

On page 134 Professor Graham suggests that the "two chronologically simultaneous episodes with contrasting results are probably as close to actual historical 'test cases' on a large order of the question of whether *different values are inherent in different theories of heredity* as our very confused and inexact world is going to present to us." (my emphasis). Further on he indicates that on the basis of the historical account he has given us, he is going to ask "what do these examples tell us about the relationships between science and sociopolitical values?" (p. 134).[1]

His answer is somewhat complex and involves what he refers to as several layers. He suggests on page 136 that the different value connotations arise from the dual nature of the genetics of the 1920s as a "pure and an applied science," and adds "whether or not science qua science contains values, most people will agree that technology does have values impacts." He then concludes in the next sentence that "therefore, the significance for values of a particular science will vary as the associated technology for that science develops."

His example to illustrate this point about technology draws a distinction between Mendelian *science* and a *technology* for assessing the value of different genotypes. He amplifies on this with a historical example, citing Ploetz's more humane suggestions to "shove selection back before birth" rather than selecting among phenotypes (i.e., living human beings).

Several pages later (p. 137) Professor Graham asserts that in his historical exposition he has "separated the question of political and social motivations from those of scientific inquiry to the maximum degree that is possible (it is never entirely possible) by focusing on the more academic debates of the twenties not the ideological exhortations of the thirties." He adds "we have separately considered the question of technology trying to show where it had differential impacts on value that are distinguishable from those of scientific theory."

Professor Graham now moves toward the main part of his thesis. The above cited separations of "academic debates" from "ideological exhortations" and of science from technology now permit him to move to his "central question . . ., one that relates to scientific theory alone: was there something inherent in a hypothetical Lamarckian theory of heredity that made it a substantial buttress for egalitarian political values, and something inherent in the Mendelian theory of heredity that lent support to elitist views of society?" (p. 138)

His answer to this central question is straightforward. He writes:

It is my opinion that in our present state of knowledge of human heredity the question of whether available theories have in themselves positive or negative value content simply cannot be answered on an abstract level, apart from reference to the existing sets of social forces and to the available technologies. However, in a given historical situation, such as the ones we have been studying here, rival scientific theories always exist within the context of given sets of social and political circumstances, of competing political ideologies, and economic motives, and of existing systems of technological capabilities. Within those frameworks of external factors, rival scientific theories *do* have differential value implications, but they derive their value meaning much more from their relationships to these external factors than from anything inherent in the science. (p. 140)

This much has been exegesis. Let me now turn to the reasons for my disagreement with the construal of the relations between science, technology, and values.

Let me begin with what I take to be a relatively minor point. I must confess that I do not see that Professor Graham has made a case for his claim that technology has any different value implications than does science. The problem may be partially because of the unclear line between science and technology or between basic science and applied science. I do, however, fail to see what value implications a more powerful electron microscope (presumably an example of technology) has *per se* than does a more powerful theory in basic science that indicates how DNA recombines. I will grant Professor Graham's thesis that important second-order links do exist between science and values, and will come back to this below, but I am dubious that an important distinction can be made between those links which are associated with technology and those which have to do with science. The differences in value implications are, I think, due to what *empirical consequences* a scientific theory has and how those consequences can be brought about by manipulation of initial test conditions associated with a theory. This however, is a situation which is dependent on the *degree of development* of a science or a technology, but usually of *both* science and technology.

Let me turn to several more important points of explicit difference that seem to separate Professor Graham and myself.

Professor Graham maintains that the second-order types of science and values interaction, which he does view as "contingent" or "relational" (p. 143), are "far stronger in terms of the concerns of historians or sociologists . . . [than are] values components that may be truly inherent in science . . ." (p. 143). These second-order interactions are, he adds, "really massive impacts of science and technology on social values" (p. 143). "These impacts that derive from the relationship of a particular scientific theory or technical innovation to a particular society in a particular historical setting . . . are the type that change history; they result in life and death, in vast changes in social mores, in changing attitudes toward the scientific enterprise itself" (p. 143).

On the other hand, Professor Graham contends that "these values that may be truly inherent in science itself"—he *is* prepared to deny the value-free character of science—"are of less significance so much so that I would call them " 'weak' " (pp. 143-44).

Now this distinction between the "strong" and "weak" interactions between science and values may be producing more heat than light. Professor Graham in introducing his senses of the terms "strong" and "weak" was reacting to comments I made on the earlier version of his essay. There I noted, as he quotes on p. 142, that his "view of the relations between science and values . . . seems to me to be an extremely *weak* one." Professor Graham suggests, I think correctly, that we are using the terms "strong" and "weak" in different senses. Since these terms have unfortunate honorific connotations, I suggest that we try dispensing with them and talk instead about "internal" and "external" relations of science and values.

The sense I would want to give to an internal relation in this context is that science *per se,* or intrinsically, contains valuational elements of a normative sort. This would deny the value-free character of science. Elsewhere I have argued for the existence of such internal values in science.[2] In terms of such internal values' *direct* impact on society, I think it would be weak. I would maintain, however, that since these internal values are both (1) what engender scientific and technological progress, in part, and (2) also are a primary source of the methodology of rational value adjudication, in that they are at the core of the scientific method of rational inquiry, such values' indirect effect on civilization is massive.

I should caution that the view I would want to defend does not require that we can obtain from science *per se* a system of values sufficient for a just society, such as Jacques Monod seemed to think in his *Chance and Necessity*.[3] Other characterizable sources of values, such as religious traditions, codes of law, intuitions of fairness, and philosophical systems of normative ethics are crucial background for developing an ethic sufficient for a humane and just society, in my view. Thus, in this sense, values internal to science have a weak aspect since they are not *per se* sufficient to generate a comprehensive ethic.[4]

The sense I would give to *external* relations of science and values is what Professor Graham refers to as his "second-order" link. Clearly, these external relations exist, but from a philosophical point of view they may be rather thin and also misdirective of our inquiry into the relations between science and values, if we focus on them too heavily. We know, for example, that there are value implications of handing a five-year-old a loaded gun. It is less clear what the value implications are of handing society the tools for cloning humans, though we suspect they will be massive on a social scale. It is unclear, however, that noting these potential massive value implications assists philosophical understanding of the full range of significant interactions between science and values. Let me amplify on this point and indicate my reasons for concern.

What troubles me about focusing on the second-order type of interaction—the "external" interactions—is that I believe it may direct our attention away from more philosophically fundamental, and also socially significant, issues associated with the foundations of ethics and its relation to science—the theme of this book. The reason why I believe this is the case is that Professor Graham in his example focuses on the value implications within a particular (or two particular) social contexts of science and technology. But it is equally important to direct attention at the valuational assumptions in those social contexts, at alternative assumptions of a valuational sort, at the ethical foundations of such valuational assumptions, at the implications produced by the scientific theories under alternative ethical assumptions, and at the methodological foundations of the adjudication process in which we examine the valuational and scientific assumptions in the light of conditions and consequences. In connection with this last point, I would maintain that it is the scientific method, construed in an appropriately general sense as critical rational inquiry, but also as particularized for specific judgmental situations, which is the most appropriate method for value-conflict clarification and adjudication.[5]

What I am suggesting then is that the significance of the external or second-order links is only a part of a broader set of relations between science and values that needs to be examined if we are to come to grips with the relations between the foundations of ethics and its relation to science.

NOTES

1. Page references given in text are to Professor Graham's essay in this volume.
2. See K.F. Schaffner, "Reduction, Reductionism, Values, and Progress in the Biomedical Sciences," in R. Colodny, ed., *Logic, Laws, and Life* (Pittsburgh: University of Pittsburgh Press, 1977), pp. 143-71.
3. J. Monod, *Chance and Necessity* (New York: Alfred A. Knopf, 1970).
4. See Schaffner, "Reduction, Reductionism . . .", pp. 166-67.
5. See J. Dewey, *The Quest for Certainty* (New York: Putnam, 1960) esp. chap. 10, and Schaffner, ibid., pp. 164-67.

Part II

The Foundations of
Ethics

Moral Autonomy

Gerald Dworkin

> The will is therefore not merely subject to the law, but is so subject that it must be considered as also making the law for itself and precisely on this account as first of all subject to the law (of which it can regard itself as the author).
>
> —Kant

> [Virtue] is not a troubling oneself about a particular and isolated morality of one's own . . . the striving for a positive morality of one's own is futile, and in its very nature impossible of attainment . . . to be moral is to live in accordance with the moral tradition of one's own country.
>
> —Hegel

1. There is a philosophical view about morality which is shared by moral philosophers as divergent as Kant, Kierkegaard, Nietzsche, Royce, Hare, Popper, Sartre, and Wolff. It is a view of the moral agent as necessarily autonomous. It is this view that I wish to understand and evaluate in this essay. I speak of a view and not a thesis because the position involves not merely a conception of autonomy but connected views about the nature of moral principles, of moral epistemology, of rationality, and of responsibility.

2. I shall begin by distinguishing a number of ways of explicating the notion of moral autonomy. In the philosophical debate very different notions have been confused, and since they are

involved in claims which range from the trivially true to the profoundly false it is essential to distinguish them.

3. The most general formulation of moral autonomy is: A person is morally autonomous if and only if his moral principles are his own. The following are more specific characterizations of what it might mean for moral principles to be one's own.

1. A person is morally autonomous if and only if he is the author of his moral principles, their originator.
2. A person is morally autonomous if and only if he chooses his moral principles.
3. A person is morally autonomous if and only if the ultimate authority or source of his moral principles is his will.
4. A person is morally autonomous if and only if he decides which moral principles to accept as binding upon him.
5. A person is morally autonomous if and only if he bears the responsibility for the moral theory he accepts and the principles he applies.
6. A person is morally autonomous if and only if he refuses to accept others as moral authorities, i.e., he does not accept without independent consideration the judgment of others as to what is morally correct.

4. In this essay I am not concerned with other issues that have been discussed under the heading of autonomy. I am not concerned with the general question of what it is for an individual to act autonomously. I am not concerned with various views that have been discussed under the heading of the autonomy of morals—whether one can derive an "ought" from an "is"; or the relations, if any, between facts and values; or whether the acceptance of moral principles necessarily carries with it a motivating influence upon conduct. Clearly, there are connections between one's views on these matters and the issue of moral autonomy. I do not propose to draw them here.

5. What could it mean to say that a person's moral principles are his own? We have already identified them as "his" when we referred to them as "a person's moral principles." But how do we make that identification? In terms of considerations such as: Which moral principles occur as part of the best explanation of a person's actions? Which moral principles would the person de-

fend as correct? Which moral principles does he use as a basis for self-criticism? For the criticism of others? Which moral principles make his enthymematic moral arguments into valid arguments? There are practical problems in making this identification—the issue of rationalization, and theoretical problems (*akrasia*—are they the person's principles if he doesn't act in accordance with them?), but there is no special problem connected with autonomy. That issue concerns the notion of moral principles being his "own."

6. How could a person's moral principles not be his own? Not by being at the same time someone else's. For the fact that we share a common set of principles no more shows them not to be my own, than our sharing a taste for chocolate shows that my taste is not my own. Perhaps I borrowed your principles, or illegitimately appropriated them, or simply found them and never bothered to acknowledge their true owner. But all of these notions (as with the idea on which they trade—property) assume the notion that is to be explained. They all assume that somebody's principles are his own and that somebody else is not in the appropriate relation to his principles that the first person is.

7. With property, how one acquired it is essential. Perhaps that is what we must look for here as well. One suggestion is that we create or invent our moral principles. Sartre speaks of a young man deciding between joining the Free French or staying with his aged mother as being "obliged to invent the law for himself." Kant says the will "must be considered as also making the law for itself."

8. If this is what moral autonomy demands, then it is impossible on both empirical and conceptual grounds. On empirical grounds this view denies our *history*. We are born in a given environment with a given set of biological endowments. We mature more slowly than other animals and are deeply influenced by parents, siblings, peers, culture, class, climate, schools, accident, genes, and the accumulated history of the species. It makes no more sense to suppose we invent the moral law for ourselves than to suppose that we invent the language we speak for ourselves.

9. This is perhaps—I doubt it—a contingent difficulty. There are logical difficulties as well. For suppose one did invent a set

of principles independently of the various influences enumerated above. What would make them *moral* principles? I may act in accordance with them and take my deviation from them as a defect but that is not enough. I might be engaged in some kind of private ritual.

10. A central feature of moral principles is their social character. By this I mean, partly, that their interpretation often bears a conventional character. What my duties are as a parent, how close a relative must be to be owed respect, what duties of aid are owed to another, how one expresses regret or respect, are to some extent relative to the understandings of a given society. In addition, moral rules often function to provide solutions to a coordination problem—a situation in which what one agent chooses to do depends upon his expectations of what other agents will do—agents whose choices are in turn dependent on what the first agent will do. Such conventions depend upon the mutual convergence of patterns of behavior. The principles of morality are also social in that they have what Hart calls an internal aspect. They provide common standards which are used as the basis of criticism, and demands for obedience. All of these preclude individual invention.

11. Does this imply that moral reform is impossible? Not at all. It just implies that moral reform takes place against a background of accepted understandings about our moral relationships with one another. And *these* are not invented. Moral reforms (almost?) always take the form of attacking inconsistencies in the accepted moral framework, refusals to extend rights and privileges that are seen as legitimate already. Analogy and precedent—the weapons of the conservatives—are the engines of reform as well.

12. If I do not and cannot make the moral law for myself, at least, so it is claimed, I can always choose to accept or reject the existing moral framework. It is up to me to decide what is morally proper. This is the proper interpretation of Sartre's claim that his young man is "obliged to invent the law for himself." Nothing in the situation he faces shows him what to do. The competing claims are equally compelling. He must simply decide.

13. Choice and decision do enter here but it is crucial to see

how late in the game they enter. For Sartre (and the young man) already know they are faced with competing claims, and that these claims are of comparable moral force. That a son has obligations to his aged mother; that a citizen has a duty to defend his country against evil men; that neither of these claims is obviously more important or weighty than the other—none of these are matters of choice or decision. Indeed, the idea that they are is incompatible with the quality of tragic choice or moral dilemma that the situation poses. For if one could just choose the moral quality of one's situation, then all the young man would have to do is choose to regard his mother's welfare as morally insignificant, or choose to regard the Nazi invasion as a good thing, or choose to regard one of these evils as much more serious than the other.

14. Could someone *choose* to regard accidental injuries as having the same moral gravity as intentional ones? Utilitarians, some of whom say something like this, do so on the basis of a *theory*.

15. Still, if one cannot originate one's moral principles, and if the relevance of various factors to moral decision making is not always a matter of choice, the ultimate weighting of the moral factors is the agent's decision and his alone. A moral agent must retain·autonomy, must make his own moral choices. The problem is to give this idea content in such a way that it escapes being trivial (who else could make my decisions?) or false (the denial of authority, tradition, and community).

16. How could a person's moral principles not be his own? Here is one case. It is from *Anna Karenina*.

> Stefan Arkadyevitch always read a liberal paper. It was not extreme in its views, but advocated those principles held by the majority of people. In spite of the fact that he was not really interested in science, or art, or politics, he strongly adhered to the same views on all such subjects as the majority and this paper in particular advocated, and changed them only when the majority changed. Or rather, it might be said, he did not change them at all—they changed themselves imperceptibly.

Here the beliefs are not his because they are borrowed; and they are borrowed without even being aware of their source; and, it is implied, Stefan is not capable of giving some account of their validity—not even an account which, say, stresses the likelihood of the majority being correct, or the necessity for moral consensus. All of these are important here—not just the borrowing

17. It is not sufficient for showing that my moral beliefs are not my own to show that my holding them has been casually influenced by others. Almost all our beliefs have been so influenced. Nor is it enough to show that while I have reasons that justify my beliefs, those reasons are not the causes of my beliefs. I may have acquired a belief from my father in, say, the importance of equality. But if I now have reasons which justify my belief, it is my belief. Nor is it enough to show that among the reasons I present to justify my belief are reasons that make reference to the beliefs of others.

18. If I believe Father knows best, and I do what Father tells me to do because I believe Father knows best, then Father's principles become mine as well. To deny this is to assume that what is mine must only be mine.

19. Underlying the notions of autonomy considered so far are assumptions about objectivity, the role of conscientiousness, obligation, responsibility, and the way in which we come to see that certain moral claims are correct. I shall argue that with respect to all of these issues the doctrine of autonomy in any of the interpretations (1-4) is one-sided and misleading.

20. These doctrines of autonomy conflict with views we hold about objectivity in morals. We believe that the answering of moral questions is a rational process not just in the sense that there are better and worse ways of going about it, but that it matters what answer we find. It makes sense to speak of someone as being mistaken or misled in his moral views. The idea of objectivity is tied up with that which is independent of will or choice. That a certain inference is valid, that a certain event causes another, that a certain course of conduct is illegal, that Bach is superior to Bachrach, that Gandhi was a better person than Hitler, that the manufacturer who substituted an inert substance for the active ingredient in Ipecac did an evil thing, are independent of my will or decision.

21. There is a paradox for notions of autonomy that rely on the agent's will or decision. Consider the statement that moral agents ought to be autonomous. Either that statement is an objec-

tively true statement or it is not. If it is, then there is at least one moral assertion whose claim to validity does not rest on its being accepted by a moral agent. If it is not, then no criticism can be made of a moral agent who refuses to accept it.

22. Another form of the paradox. Consider the following two claims.

1. P ought to be autonomous
2. P chooses not to be autonomous

Does P have any reason to accept (1)?

23. We can see in Kant the confusion engendered by his attempt to reconcile objectivity and autonomy. For Kant the moral law does *not* obtain its objective character by being chosen or willed by us. The categorical imperative commands us to act on that maxim which we *can* will as universal law. In a second formulation we are enjoined to act *as if* the maxim of our action were to become through our will a universal law of nature. What is essential is that one could will to act in such and such a way, not that we actually so will.

24. But when Kant faces the problem of how such an imperative can be binding on us he reverts to the notion of willing. The argument is that a categorical imperative cannot be binding because of some interest I have—because then it would be hypothetical. So, in the philosophical move than Putnam calls the "what else argument," Kant concludes that it must be binding because we have legislated it ourselves. But there are other possibilities, including the thesis that there are objective requirements of reason which provide their own form of rational motivation.

25. For Rawls, the objectivity of principles is defined in terms of their being the principles that would be chosen by free, equal, and rational beings. As such they are binding upon individuals whether or not they view them as binding. But agents are able to put themselves in the position of being choosers, to follow the arguments for the principles, and to desire that everyone (including themselves) accept such principles as binding. To the extent that they are motivated in this manner—and not, say, by mere submission to parental authority—they are morally autonomous.[1]

26. In addition, there arises at the level of the interpretation

and application of moral principles a substantive political conception of autonomy. Given the fact that real persons, even if they accepted a common moral framework, will have different and conflicting ideas of the correct interpretation of that theory, a state may be required to recognize political autonomy of its citizens. That is, it may not restrict the liberty of individuals unless it can justify such restrictions with arguments that the person himself can (given certain minimal rationality) see as correct. Such a doctrine can be invoked in defense of freedom of expression and conscience. But it is important to note that this argument applies to a specific area of moral life—the limits of state power. And even there, I have argued elsewhere, there are difficulties.[2]

27. Preoccupation with autonomy carries with it the attribution of supremacy to the concepts of integrity and conscientiousness. For if what is morally correct is what one has decided for oneself is correct, then for another to interfere with one's freedom of action based on that decision is to stifle moral personality and encourage hypocrisy. Politically this leads to the type of philosophical anarchism that Wolff espouses.[3] Socially it leads to the rejection of any use of community and peer pressure to limit the liberty of individuals. This is often defended in the name of Mill—a defense that only charity could attribute to a misreading (since it is obvious that anyone who believes this cannot have read Mill at all).

28. At the very least a defender of autonomy must distinguish between autonomy of judgment and autonomy of action. The arguments in favor of allowing people to determine for themselves what is right are more compelling than those that favor allowing people to always act in accordance with their beliefs. It is one thing to argue in defense of freedom of expression from premises concerning what individuals who wish to retain autonomy would be willing to grant the state by way of authority to limit expression. It is quite another to argue, as Wolff does, that law qua law creates no obligation to obey.

29. As for integrity—that second-order value which consists in acting in accordance with one's first-order values—it is not to be despised. There is something admirable about the person who acts on principle, even if his principles are awful. But there is

also something to be said for Huck Finn, who "knowing" that slavery was right, and believing that he was morally damned if he helped Jim to escape, was willing to sacrifice his integrity in favor of his humanitarian impulses.

30. A moral theory that stresses the supremacy of autonomy will have difficulties with the concept of obligation. As the etymology suggests, to be obliged is to be bound. And to be bound is to have one's will restricted; to have one's moral status altered so that it is no longer one's choice how one should act. The usual suggestion of the defender of autonomy is that one can, so to speak, tie oneself up. And this is the only way one becomes bound. All obligation is ultimately self-imposed, hence a product of one's decision or choice to be so limited.

> I am persuaded that moral obligations, strictly so-called, arise from freely chosen contractual commitments between or among rational agents who have entered into some continuing and organized interaction with one another.[4]

(How the notions of autonomy, individualism, Protestantism, and contract emerge (and merge) in moral, social, religious, and economic thought is a subject (still) worth historical investigation.)

31. This attempted solution cannot succeed. Tying oneself up is binding only if the knot is no longer in one's hands. For if I can, at will, release myself I am only in appearance bound. As Hume put it with respect to the obligations created by promises, on this view the will

> has no object to which it could tend but must return to itself *in infinitum*.[5]

To say that promises create obligations because they create expectations is true enough but of no help here. For that these expectations have moral weight is not itself chosen or decided by the maker of the promise.

32. Another way of looking at this. From the temporal perspective the commitments of my earlier self must bind (to some degree) my later self. It cannot always be open for the later self to renounce the commitments of the earlier self. This implies that even self-imposed obligations create a world of "otherness"—a world which is independent of my (current) will and which is not

subject to my choices and decisions. The distance between my earlier and later selves is only quantitatively different from that between myself and others.

33. In his discussion of the one state he believes has authority and is consistent with autonomy—unanimous direct democracy—Wolff fails to see this point. He argues that there is no sacrifice of autonomy because all laws are accepted by every citizen. But this is only true at a given point in time. What if the individual changes his mind about the wisdom or goodness of the law? Is he then bound to obey it?

34. Leaving the internal difficulties to the side, the claim that all obligations are self-imposed does not fit the moral facts. That I have obligations of gratitude to my aged parents, of aid to the stranger attacked by thieves, of obedience to the laws of a democratic and just state, of rectification to those treated unjustly by my ancestors or nation are matters that are independent of my voluntary commitments.

35. The defender of autonomy has a particular picture of the role of discovery in morality.

> [the moral agent] may learn from others about his moral obliga-
> tions, but only in the sense that a mathematician learns from other
> mathematicians—namely by hearing from them arguments whose
> validity he recognizes even though he did not think of them
> himself. He does not learn in the sense that one learns from an
> explorer, by accepting as true his accounts of things one cannot
> see for oneself.[6]

This picture is inaccurate even for the mathematician. Mathematicians often accept results on the word of other mathematicians without going through the proofs for themselves. And they may do so (particularly) in fields in which they do not possess the techniques to assess the proof even if they were inclined to do so.

36. The image of the explorer is an interesting one and is analogous to the role of the seer in the moral systems of various tribal peoples. Lest one think that this view is one that only "primitives" hold, compare the role of the "practically wise" man in Aristotle.

37. For Aristotle moral virtue is a disposition to choose which is developed in the process of choosing. We do not do good acts

because we are already good (at first anyway). We do good acts, and in doing so become good. This is paradoxical. How are we to identify those acts which are good, if we are not ourselves already good? By aiming at the mean which is determined by the "proper rule." How do we identify the proper rule? It is "the rule by which a practically wise man would determine it." Thus, to be morally virtuous one must follow the example or precept of one who is practically wise.

38. Such an account, which places reliance on the exemplary individual, on imitation of goodness, on what would in a more barbarous term be called role-modeling, seems to me to be, if not the whole story, at least a significant part of it. Such a view has its own vices of excess. There is, no doubt, the morally unappetizing sight of the person who abandons all attempts to think critically about who he is imitating, who imitates out of laziness or fear or sycophancy. This excess has received its share of attention from an excessively protestant and individualistic age. I am calling attention to the opposite defect. The refusal to acknowledge the very idea of moral authority, the equation of imitation with animal characteristics (copycat; monkey see, monkey do), the identification of maturity with doing things without help, by (and for?) oneself.

39. Consider the fifth interpretation of autonomy—being responsible for the moral theory we believe correct and for the interpretation of the principles that follow from it. Leaving aside general metaphysical doubts about freedom of the will and empirical doubts about causes of our conduct which are beyond our control, this thesis seems to me correct but vacuous. One cannot evade responsibility by asserting that one was only following orders or doing what everybody does or accepting the general will. But this leaves completely open the issue of whether one ought to be autonomous in the sense of (6), i.e., being prepared to examine one's principles in a critical fashion. Whether one does so or not, one will still be (held) responsible.

It is the confusion of these two distinct notions that leads Aiken to assert that

> no man is morally responsible for actions unless they are performed for the sake of principles which he cannot in conscience disavow.[7]

This implies that all one has to do to avoid responsibility is either be completely unprincipled or accept principles without conscientious scrutiny.

40. We come now to the last interpretation of autonomy. A person is morally autonomous only if he

> cannot accept without independent consideration the judgment of others as to what he should do. If he relies on the judgment of others he must be prepared to advance independent reasons for thinking their judgment likely to be correct.[8]

41. This is the denial of any strong notion of moral authority. On this view none of the following justifications could be acceptable.

> These principles are acceptable because they are the revealed word of God.
>
> These principles are acceptable because they are part of the moral tradition to which I belong.
>
> These principles are acceptable because the elders have pronounced them acceptable.
>
> These principles are acceptable because they are those of my class or my clan or my comrades.
>
> These principles are acceptable because they are embodied in the common law or the Constitution.
>
> These principles are acceptable because they were passed on to me as part of my training as doctor, lawyer, Indian chief.
>
> These principles are acceptable because they are customary, or the ways of my father and my father's father.
>
> These principles are acceptable because they are in accord with Nature or the Tao or the course of evolution.
>
> These principles are acceptable because they are those of Gandhi or Thoreau or Socrates or Confucius or Jesus or Tolstoy.

42. The idea of there being independent reasons for thinking the judgments of an authority correct is ambiguous. There are reasons for thinking his *judgment* likely to be correct, i.e., independent reasons for believing the content of his judgment correct. Or there can be independent reasons for thinking *his* judgment to be correct, i.e., for thinking *him* likely to be right. Corresponding to these we have a weaker and stronger notion of "checking"

this is a corruption which every disposition recognized as a virtue is apt to suffer at the hands of fanatics.[11]

49. I have argued that there is no interesting thesis about moral autonomy which follows from any conceptual thesis about the nature of morality or moral agency. It is a substantive thesis and represents a particular conception of morality—one that, among other features, places a heavy emphasis on rules and principles rather than virtues and practices. Considered purely internally there are conceptual, moral, and empirical difficulties in defining and elaborating a conception of autonomy which is coherent and provides us with an ideal worthy of pursuit.

50. It is only through a more adequate understanding of notions such as tradition, authority, commitment, and loyalty, and of the forms of human community in which these have their roots, that we shall be able to develop a conception of autonomy free from paradox and worthy of admiration.

NOTES

1. John Rawls, *A Theory of Justice* (Cambridge, Mass.: Harvard University Press, 1971), pp. 516-19.
2. Gerald Dworkin, "Non-neutral Principles," *Journal of Philosophy,* vol. 71, no. 14 (August 15, 1974).
3. Robert Paul Wolff, *In Defense of Anarchism* (New York: Harper and Row, 1970).
4. Robert Paul Wolff, *The Autonomy of Reason* (New York: Harper and Row, 1973), p. 219.
5. David Hume, *Treatise of Human Nature* (Oxford: Clarendon Press, 1888) pp. 517-18.
6. Wolff, *In Defense of Anarchism,* p. 7.
7. Henry Aiken, *Reason and Conduct* (New York: Alfred A. Knopf, 1962), p. 143.
8. Thomas Scanlon, "A Theory of Freedom of Expression," *Philosophy and Public Affairs,* 1 (Winter, 1972): 216. This is not put forward by Scanlon as a notion of *moral* autonomy. I adopt his words for my own purposes.
9. Aiken, *Reason and Conduct,* p. 143.

10. R. M. Hare, *Freedom and Reason* (Oxford: Clarendon Press, 1963), p. 2.
11. Michael Oakeshott, *On Human Conduct* (Oxford: Clarendon Press, 1975), p. 238.

7

The Rights of Persons in Plato's Conception of the Foundations of Justice*

Gregory Vlastos

IN BOOKS II TO VII of the *Republic* (hereafter "*R*.") Plato under-takes to do something never previously attempted in the history of the West: to determine on purely rational grounds all of the rights which all of the members of a particular society ought to have. The society he has in view is a Greek polis. He does not question the restrictions on its membership—the exclusion from it of all the persons in its territory except those whose title to civic status is hereditary.[1] So he ignores two large classes of persons whose presence in his native city had been for generations indis-pensable to its economic existence: resident aliens and slaves.[2] Of neither group is it stated that it is to exist in the ideal state of the *R*. To resident aliens there is no reference. To slaves there is just one—but it is not so unambiguous as to put beyond controversy the question of whether or not this institution is being retained when so much else is being discarded in the remaking of the polis.[3] I have myself participated in this controversy, arguing for the affirmative.[4] In this chapter I shall follow Plato's example and ignore aliens and slaves in discussing his theory. But I can hardly do so without at least calling attention, in passing, to one of the implications of Plato's willingness to blanket in silence the very existence of two groups of human beings whose lives were intertwined with his throughout his life. From just this fact we can infer that the idea of *human* (or "natural") rights—rights which, in Locke's phrase, "belong to men as men and not as members of society"[5]—is lacking in Plato's conception of morality. It is so alien to its basic assumptions that he does not even allude to it—to say nothing of undertaking to argue against it, explaining why he finds it inconsistent with his concept of justice.

How does Plato propose to ascertain the rights of those who are to be the members of his polis? By a purely a priori method: Socrates undertakes to discover the nature of justice by constructing a "perfectly good" (427E) polis. To be perfectly good it will have to be, among other things, perfectly just. So if we don't know what justice is, we can find out by inspecting the finished construction. The structure of rights and duties exhibited in this perfectly just society will tell us what is the just allocation of rights and duties in a polis.

Is this procedure epistemically viable? If we don't know what justice is to begin with, how can we tell what to put into our construction to get a perfectly just polis out of it at the end? To discover justice by this method would we not have to know the very thing we are looking for—before we found it? Plato would not be unprepared for this question. He must know that his proposal will look paradoxical. Earlier, in the *Meno,* he had confronted what he calls there the "eristic argument":

> It is humanly impossible to go looking either for what one knows or for what one does not know. One would not go looking, surely, for what one does know—there would be no need of that, if one already knows it—nor yet for what one does not know—for then one would not know what to look for (80E).

Plato's answer to this paradox is his doctrine that "learning is recollecting"—that all discovery is a matter of reawakening in the soul knowledge it once possessed in a primordial past and can now recover, if it will only make the right use of what is still left in it of that initial deposit (81A-D). This, he implies, is what happens in the interrogation of the slave boy (92A-85B). The question put at the start—"What is the side of a square whose area is twice that of a given square?"—baffles the boy. But his mind is not a blank. Though he has never learned geometry, he has some understanding of what "square" means, can recognize squares, can tell some of their properties, and can reason his way to other properties they are bound to have if they already have the ones which make them recognizably square. Let him proceed in step by step inference from those simple truths which are already in his mind in the form of scattered true beliefs, and he will wind up sooner or later discovering the answer to the initial question.

Though Plato does not allude in the *R*. to the doctrine of recollection and makes no comment on the epistemic side of the inquiry in Books II to V, there is no reason to doubt that the answer to the objection would be the same. At the start of the inquiry we don't know what we shall come to know at the end; but this does not preclude our having sundry true beliefs about just dealings between man and man—beliefs that Socrates can bring in by bits and pieces, as the inquiry proceeds, and weave into a coherent pattern which will exhibit the social structure of the perfectly just polis. These beliefs are not announced as principles of justice. They are introduced as principles of rational action whose intuitive appeal is meant to be so high that none of the interlocutors will wish to challenge them—as they in fact do not, taking them to be the plain common sense of social conduct with which everyone would agree. But there is no doubt that they are meant to be valid principles of justice, else how could Socrates have expected them to serve him as stepping stones to the final result: insight into the nature of justice?

The construction starts with a settlement that Socrates chooses to call a polis (369C4, etc.) though it is only a simple form of economic association. Falling in with this fiction, I shall call this primitive make-believe polis, "protopolis." Its people had to choose between two options: on one hand, self-sufficiency, every man working only for himself, relying on his own labor to meet all of his needs; on the other, interdependence, every man working for himself *and* for each of his neighbors, by specializing in the production of one of the commodities which both he and they need, and exchanging the product against those produced by others, each of whose labor has been similarly specialized. In choosing the second option, the people of protopolis could be viewed as forming a mutual benefit association: by dividing the labor and exchanging its product everyone's efficiency is enhanced, the common standard of living is raised, so everyone's work benefits everyone, including himself. That Plato so views this pattern of economic life is reasonably clear from his description of it: He speaks of the participants as cooperators, not competitors—as "partners and helpers" (369C), each of whom "places his own work at the disposal of all in common" (369E, in Lindsay's translation).[6]

Plato can view this process in this way because he does not see the shift from the first to the second option as a shift from production for use to production for the market and therewith to production for profit. For Plato both options involve production for use: for one's own needs only, in the first; for the needs of all, in the second. His description gives no place to production for an abstract, impersonal, market, where sellers compete for buyers, their competition providing a pricing mechanism for their products, and pitting producers against one another in a contest which punishes the less efficient by driving them out and rewards the more efficient by offering them a bigger share of the market, thereby allowing them to enlarge the scale of their operations and transform themselves from direct producers into entrepreneurs. As Marx points out in a perceptive comment on our passage, Plato's point of view and that of other classical authors (he quotes also Thucydides and Xenophon) differ profoundly from that of "modern political economy" whose spokesmen (Adam Smith & Co.) see in the division of labor "only the means of producing more commodities with a given quantity of labor and, consequently, of cheapening commodities and hurrying on the accumulation of capital."[7] In Plato's picture of the economy of protopolis, capital does not appear as such. The making of agricultural and industrial tools is noticed (307C-D) but is not described as production of capital goods. There is nothing here answering to entrepreneurial activity by capitalists making investments, i.e., foregoing consumption for the specific purpose of devoting wealth to operations designed to increase it. Money functions only as a medium of exchange, and the merchant only as a middleman (317B-D).[8] Only later on, when protopolis, with its healthy simplicity, is succeeded by the luxurious "fevered" city, do men "turn to boundless money-making, over-passing the bounds of the needful" (373D).

To be sure, one thing is said in the account of protopolis which may suggest the profit-seeking which is the *primum mobile* of a capitalist economy. Socrates asks:

> And in the give-and-take, whether one gives or takes, does not one act in the belief that this would be better for oneself? (369C6-7).

In the Cross-Woozley commentary, Plato is seen as maintaining here that these people are acting "entirely selfishly."[9] I submit that this interpretation, though plausible, is unwarranted. Everyone's seeing that cooperative reciprocity is "better for himself"—better than isolationist self-sufficiency—does not itself preclude his seeing also that it is better for the others too, the enhanced benefit to himself being so connected with enhanced benefit to them that neither could occur without the other. Why then should each one's motivation be *entirely* self-regarding? Why should it not be self-regarding and other-regarding too?[10] There is no reason why it should not. Why then does not Plato say so? Why should he say only "because it is better for himself"?[11] A fair guess would be: Because he does not want to get ahead of himself in his construction. The time will come when he will want to tell us that the partners in a mutual benefit association will have a fraternal feeling for each other.[12] He may be already hinting at something of the kind even at this stage of the construction, when he speaks of these people a little later (327B) as "living in pleasant fellowship with each other."[13] But that this will be their attitude toward each other is not *entailed* by the pattern of interdependent economic activity he has laid down. The norm of just social relations remains underdetermined at this point.

It also remains underdetermined at another, not unrelated, point: no provision has yet been made to indicate at what ratios the producers would exchange their products. What is there, then, to preclude gross inequalities in these ratios which would be hard to reconcile with one's intuitive sense of justice, be this ours or that of the Greeks, for, as I pointed out in *SJR* (18-19),[14] the linguistic bond of justice with equality was even closer for the Greeks than it is for us: *to ison, isotēs* ("the equal", "equality") would be the very words to which they would turn for a natural, unstrained, one-word variant for *to dikaion, dikaiosynē* ("the just", "justice") . . . (It) leads Aristotle to declare (*Nicomachean Ethics* 1130B9) that "justice *is* equality, as all men think it, even apart from argument."

Let me suggest a scenario. Suppose that one man in this settlement has a natural monopoly on some skill: he can make fine wooden flutes, an article which is much in demand in this community and can only be made by someone who has an ear with "perfect pitch" and manual skill to match; only this fellow has this capacity, and being a fast worker he can turn out a flute in four hours. So working four hours every fourth day, turning out ninety flutes per annum, and charging in exchange a share of the goods produced by others which is twice as large as that of everyone else, his position relative to all the others is enviable indeed: working one-eighth of the time the others do, he gets twice as much as they; one hour of his working time is being paid as much as sixteen hours of theirs. Is there anything in Plato's description of the protopolis to rule out this kind of inequality? Nothing, so far as I can see. Assuming that those people prize a perfectly pitched flute as you and I might value a fine antique or a Picasso, why shouldn't each feel that it is "better for himself" to have this character in their group, in spite of his getting away with so little work for so large an income? Aren't all the others better off with him in the group than they would be without him?[15] So what is wrong with his being so much better off than any of them? If it grates on our sense of justice, why should it?[16] May it not be only vicarious envy that makes us feel morally uneasy about his ability to command terms so privileged by comparison with those of his peers? Would Plato's theory deny the justice of such an arrangement and, if so, on what grounds?

It would indeed, but not on the ground of anything mentioned or implied in the account of the protopolis. To find it we must move on, though even so we shall not get it in the form of a direct statement. We shall have to derive it from the decisions governed by a new principle which is never spelled out. Let me take it upon myself to formulate it on Plato's behalf so that we can have it before us as clearly and explicitly as possible:

> All members of the polis have equal right to those and only those benefits which are required for the optimal performance of their function in the polis.

I shall call this the Principle of Functional Reciprocity. For "function" I could have used "service" or "work." Plato uses various words for it.[17] Terminological rigidity is not to his taste. Regardless of its verbal vehicle this is the key component of his concept of the justice of the polis, as is made abundantly clear at various points. Thus it turns up in the final definition of the justice of the polis which is meant to crystallize the cumulative insight gained in the preceding inquiry. The abbreviated formula at 433B, "doing one's own," is a contraction for "doing one's own work" or "performing one's own function." In its expanded form in the preceding sentence (433A5-6) it reads (in one of the possible translations),

> each single person pursuing that single practice pertaining to the polis which his own nature is best fitted to pursue.

A still more complete description of it, expanded to include additional items introduced at various points in the preceding account, might run:

> A politically assigned life-long work role—for which one qualifies by natural aptitude, by politically provided education, and by excellence of performance—which constitutes the unique vehicle for one's optimal contribution to the happiness and excellence of the polis.[18]

To comment on each of the items in this description would call for another essay. The two things I would emphasize the most if I were doing that job would be, on one hand, the terms "politically" and "polis," and on the other, "excellence of performance." The first explicitly identifies function in reference to the polis, no longer now a euphemism for the purely economic association with which we started in the protopolis, but a full-blooded state, and indeed much more than that by modern notions of the state—not only complete with apparatus of force and law, but bloated to include every single other public institution, educational, cultural, and religious, and exercising control over the whole life of the citizen, including his private life and that of his family, if he is allowed to have one. The second is best conveyed by the alternative term Plato sometimes uses to describe the citizen's function: *technē*, "craft,"[19] conceived in the typical Greek way, which thinks of a craft *as* an "art"—the other sense of *technē*—placing as high a premium on the beauty as on the utility of what a craft creates,[20] expecting from the craftsman something of that striving for perfection which had produced in Athens such marvels of workmanship by common craftsmen as the curvature of the stylobate and entasis of the columns of the Parthenon, or those sixth—and fifth—century vases, some of them made by slaves, which are now counted among the world's masterpieces of design. Plato extends this conception, making it cover not only the craftsman's product but also the character of the craftsman himself: he is expected to make of his own self also something beautiful, as well as useful, for the polis.[21]

But I cannot afford to expand on this. I must content myself with a remark on the only immediately relevant feature of this function: the one which ties it directly to the concept of justice and of rights: One's function, as Plato conceives of it, is both the citizen's master-duty and, at the same time, his master-right. The first is emphasized both in the "doing one's own" formula, where "doing" clearly refers to a duty, not to a right, and in sundry ad hoc statements, as when it is said of the guardians that their function is to constitute the "one goal in life with a view to which they do everything they do, both in private and in public" (519C), and that they "must practice nothing which does not conduce" to this single goal (395B-C). But this is only one side of the coin. The other is that the discharge of this function is for each of these people a privilege, an infinitely precious one, the basis of the worth and meaning of their life, so much so that if through some calamity they were to become unfit to do their work, life would lose its value for them: if they cannot do their function, they would just as soon be dead. Socrates brings this thought into his construction by a digression on the culture-hero Asclepius, boldly reconditioned by Plato's work-ethic to believe that it was not right to give (medical) treatment to a man who could not live in his established round of duties; this would not be profitable either for the man himself or the polis (407D-E).

The implication is that the polis would not grant him the right and, if he were self-respecting, he would not claim it. A carpenter, offered extended house-bound treatment by an unpolitical doctor, turns it down, replying that "life is not worth living in that way, keeping his mind on his illness and neglecting his proper work." So he goes back to his job where he either regains his health or "if his body will not stand the strain, solves his troubles by dying" (406D).

So much for the term "functional" in my label for Plato's principle. What of the other term, "reciprocity"? This reaffirms the conception of the polis as a mutual benefit association where everyone, by performing his own function, benefits everyone in the polis, counting on everyone else to benefit him along with the rest as each performs his proper task. And this conception is tightened up far beyond its first appearance in protopolis. The *only* benefits one is now allowed of right from the cooperative interchange are those which are required for the optimal performance of one's function. What difference this makes will be made clear shortly. But first I must ask how reciprocity, thus maintained and intensified, will relate to equality[22] on whose intimate connection with the idea of justice in Greek speech and thought I remarked above. To press this question on Plato's text is to become aware of a remarkable fact to which I drew attention in *SJR*:[23] that the words "equality" and "equal" are consistently avoided throughout the whole of the construction which culminates in the definition of justice in the middle of Book IV, and then resumes in Book V to produce there a ringing manifesto of equality between the sexes within the guardian class, without a single use of the word "equality," without so much as an allusion to it. From the philologist's point of view, who is concerned at least as much with words as with thoughts, and with words avoided no less than with those used and overused, this is, at first sight, startling. It is unparalleled in the whole of the Platonic corpus or in that of any other Greek philosopher, to my knowledge. How could Plato have wished to freeze out of his discussion of justice a word which functioned in his mother tongue as a virtual synonym of "justice"?

A part of the answer is that equality was too much of a political hot potato. For Plato's public the question concerning the rights of persons whose urgency remained paramount over that of all other public issues concerned the just allocation of political rights which, for the Greeks, meant the right of direct participation in functions of government and therewith a share in the control of the state. Democracy and oligarchy were the main options—the only ones, where tyranny had not preempted them. Equality was the democratic watchword. This, needless to say, was not all-around equality: the redivision of property, the liberation of women, the abolition of slavery were never seriously at issue in the debate. Democracy stood for substantive equality in the political sphere: the equal right to sit in the Assembly and, through the lot, the equal chance to serve on the Council, the discasteries, and on one or more of the multitudinous magistracies. Oligarchy stood for the contrary, that is to say, in plain language, for political *in*equality. But given the ancient ties of justice with equality in the language, conservative critics of democracy and advocates of oligarchy could hardly make "inequality" their battle cry. So the notion of "proportional equality"[24] was evolved—a brilliant piece of verbal *legerdemain,* keeping the word while denying its substance. "Equalizing" rights by proportioning benefits to merit would call for any degree of inequality that could plausibly be thought to comport with unequal merit; so the propertied classes, with the higher merit they could claim for themselves on the strength of their wealth and breeding, would find it natural to believe that they, and they alone, had the right to participate in the effective control of the state.

Where would Plato stand in this debate? Since in the *R.* he puts oligarchy above democracy in the gradient of political degradations,[25] he makes it clear enough that if he had to choose between oligarchy and democracy in the political arena, his choice would be for oligarchy, and that he would prefer the still more restricted form of oligarchy he calls timocracy—a constitution of the Spartan type. But there could hardly be a graver misunderstanding of his conception of the justice of the polis in the R. than to see it as a defense of oligarchy against democracy. Plato had an unusually clear-eyed view of the economic basis of this political conflict. He sees it as, at bottom (and all too frequently on the surface, too), a conflict between the "two poleis," "the city of the poor, and the city of the rich" (422E-423A).[26] And he regards this very division as the disease of the city-state, chronic, endemic, and certain to prove fatal unless a cure is found. The cure is to be his functional state, where classes are articulated along functional lines, and the social differentiations are not divisive but integrative. How then could he think that the justice of this polis could be expressed in terms of the slogans of either of the contending parties? That is why, perhaps, Plato shies away from the word "equality" in his construction of the ideally just polis. That he would think the democrats' "arithmetical" equality anathema is only too clear. Could he, then, content himself with its oligarchic counter-slogan, "proportional equality?" How could he, when he thinks the oligarchic constitution a disease, less painful than democracy, but no less of a mortal illness of the body politic for all that? This, I think, is a possible explanation of why this question is never raised in Books II to V of the R., and why Plato walks so gingerly around all kinds of topics that would provoke it.

But there are some questions one cannot avoid, no matter how hard one may try. Equality is such a one for a philosopher who has undertaken to determine what an ideally just polis would be like. Sooner or later he must confront the question whether or not his society can be ideally just unless in some important ways it proves ideally equal in the full sense of that term, i.e., unless its members have the *right to derive equal benefits* from its cooperative give-and-take and also the *right to an equal voice* in decisions affecting the direction of their joint activity.[27] In opting for functional reciprocity, Plato emphatically rejects equality in this ideally full sense; if the members of the polis are to have only those rights which are required for the optimal performance of their respective function, their unequal capacities will dictate unequal rights to share in the distribution of the goods produced and in the governance of their common life. Yet even so, the principle of functional reciprocity, as used by Plato, while precluding equality in the sense defined, will still operate as an effective constraint on permissible inequalities, blocking those for which a functional justification cannot be found. In this way functional reciprocity does work for Plato of a radically different nature from what the principle of "proportional equality" would have done. On the surface the two principles may seem very similar. But when put to work they give different results. Functional reciprocity will block inequalities which would have passed through proportional equality like a sieve. Let me illustrate by harking back to the flute-maker in protopolis discussed above.

To justify by the principle of functional reciprocity (hereafter "FR") those handsome privileges he has carved out for himself, he would have to make a case that they are required for top performance at his job. It is hard to see how he could do that. Is there any good reason to think that he can't make as many flutes of as good quality unless he is fed, clothed, shod, housed, etc., so much better than the rest of the folk? No. Or unless his work load is such a tiny fraction of theirs? No. So he must climb down from his high horse and put in a longer working day at lower real wages. If no more flutes are needed, there are plenty of things a man with his fine ear can do to enrich the services supplied to the community. He might, of course, refuse, threatening slowdown or work stoppage unless he is given those fancy terms. But this would be construed as flagrant violation of his master-duty, and a community run on Platonic lines would rather do without flutes than tolerate a moral viper in its bosom. So the FR principle has a good cutting edge. Would the principle of proportional equality (hereafter "PE") serve as well? So far as I can see, it would be of no use at all. Why shouldn't the flutemaker claim that the merit of his superior talent is so much greater than of mere shoemakers and the like that it entitles him to vastly greater social benefits? The notion of merit is so soft, it is impossible to see how claims for the superlative merit of his work could be refuted in the face of the fact that he, and no one else, can make those delectable flutes for which the other workers are willing to exchange such big chunks of their earnings, however bitterly they may resent his extortionate terms.

Let me consider now decisions made by Plato himself concerning rights of persons. I want to show how the FR principle, though never formally stated, is in fact used. I pick out for this purpose two such decisions which affect profoundly the whole design of the ideal polis. I shall argue that both are understandable only as applications of the FR principle: the PE principle would not have led to the same results. The examples are the rights of women within the guardian class, and the rights of the guardian class taken as a whole against those of the producers.

If we were to mean by "feminism"—as we could without eccentricity—the view that the rights of women should be equal in every way to those of men, then Plato must be counted as the first Western philosopher to implement a radically feminist position within an important segment of society; and not only the first, but, for more than two thousand years, the only one.[28] To find another major philosopher who qualifies we would have to go to John Stuart Mill. To be sure, Plato's feminism was only for the elite. How women are to fare among the producers in his state remains obscure. Since the nuclear family survives in their case, some (at least) of the disabilities of women that had been built into that institution would undoubtedly remain. But that is no reason for failing to give proper recognition to the fact that in the part of his society which Plato thought all-important—where all the creative thinking, planning and deciding would be done and where all the power would be concentrated—there women would be admitted on a par with men and treated as the absolute equals of men.[29]

The constrast between the status of the women guardians in Plato's polis and that of the women in contemporary Athens—of all the women there, but especially the middle- and upper-class women, on whom the segregationist constraints bore down most heavily of all—is polar. Here is a rundown of the main points in a vis à vis comparison:

1. Right to education. Schools, gymnasia, palaestrae, a male monopoly in Athens, are open without distinction to both sexes among Plato's guardians.
2. Right to vocational opportunity. Above the proletarian level, none in Athens. In Plato, career is open to talent of both sexes on equal terms.
3. Right to unimpeded social intercourse. Except at the proletarian level, the Athenian woman is born, lives, and dies

in a domestic ghetto. In Plato her living space is the same as man's in the upper two classes.

4. Right to heterosexual choice. This is grossly unequal as between men and women in Athens, both before and in marriage. Among Plato's guardians it is strictly equal: equally restricted for men and women within the reproductive age, equally unrestricted for both thereafter.

5. Legal rights. In Athens a woman is the legal ward of her nearest male relative. In Plato a woman guardian would be as much of a legal person as would a man.

6. Political rights. None for women in Athens, the same for women as for men among Plato's guardians.

What took Plato to this position? Certainly, nothing personal—there was no Harriet Taylor in Plato's life—nor yet any kind of compassionate response to the deprivations that were the daily lot of women: when he refers to the deplorable effect which social, cultural, and vocational segregation has on women's intellect and character in Athens, there is no gleam of sympathy for the victims—only regret for what is lost to the polis by the failure of half its adult members to realize their intellectual and moral potential in its service.[30] As is clear from the text of Book V of the *R.*, Plato's feminism is a direct result of his theory of justice and, more particularly, of that component of the theory which answers to the FR principle. That membership in the guardian class should be as open to women as to men is argued for solely on the ground that biological femaleness is irrelevant for good performance in any recognized work role.[31] Though Plato believes that within any given function women are somewhat inferior to men on the average, he reckons the degree and distribution of that difference to be such as to enable many women to qualify for admission to all of the higher work roles and for performance there at a level superior to that of many men. Could he have got the same result from PE? We can see what happens to the rights of women in the *Laws* where PE is hailed as *the* principle of justice (757C) and FR is ignored. When that occurs women's rights take a beating. The gains they had made in the *R.* in respect of educational opportunity and unimpeded social intercourse they retain. But they lose their gains on three other fronts: legally, politically, and vocationally they are virtually back in their traditional inferior Athenian status.[32] How exactly Plato would have thought that the justice of these inequalities is validated by PE is unclear. What is clear is that this principle is so vague and loose that it can offer no resistance to them, while if FR were being applied it would have ruled them out as firmly here as previously in the *R*. Plato could hardly have shed his belief that some women are better than many men at many tasks. How then could he have denied them access to all except household tasks consistently with the FR principle?

Now, lastly, a quick look at what is, after all, the most important single decision concerning the rights of persons in the ideally just polis of the *R*: that the great bulk of the civic body are to have no political rights—no right to participate in any of the deliberative, legislative, judicial, or executive functions of government: these are to be reserved to a tiny elite. Since the Platonic polis possesses its members body and soul—their mind, sensibility, religious feeling, conscience—the power over the masses which is now vested of right in the governing group is immense. A more extreme inequality in the tenure of political power has never been conjured up by the Greek imagination. And it is justified solely by the claim that governing is a function whose optimal performance Plato thinks only philosophers can achieve.[33] Could it then be seriously argued that even here, in what has all the earmarks of a colossal rip-off, the FR principle nonetheless supplies a constraint on the ensuing inequalities? It can. Those vast powers that are accorded the philosophers they must earn by submitting to harsh institutional restrictions on their personal life, designed to block any nonfunctional, self-indulgent use of their powers. Hence private property and private family are denied them—deprivations whose severeity has no precedent in the regimen of any ruling class recorded in Greek history or pictured in Greek fantasy. Moreover, they are required to make large sacrifices of another sort. For fifteen years they are required to give up from their studies as much time as will be needed to carry the burden of the day-to-day administration, instead of having the bulk of it done for them by subordinates co-opted for this purpose from the lower orders.[34] The rationale of these constraints is clearly dictated by FR, while contrariwise it could not be by PE: there is nothing in the notion of "merit" to prescribe similar or equivalent constraints.

Nor is it clear how PE could have sustained the sense of affectionate partnership between the powerful elite and the powerless mass which Plato is counting on to provide the morale of the just society. He refers repeatedly to this in the *R.*, but most revealingly in that passage near the end of Book IX (519C-D) which I discussed at length in *SJR.*[35] The producers' total exclusion from any right to share in the government is acknowledged in a word the Greek reader would have thought shocking to see applied to freemen, even if only metaphorically, and with benign intent: *doulos* (slave). Affirming in this way the producer's total subjection to the philosopher, Plato still insists that the relation is beneficent and will be felt as such, for through it the two will become "alike" (*homoioi*),[36] and *philoi. Philoi* I rendered "friends," but only for want of a stronger word that does not go all the way to "lovers." Much earlier, in the Myth of Metals (414 ff.), Plato had said: "All of you are brothers in the polis" (415A).[37] Fraternal solidarity is what he has in view, affection generated in a working relationship where the sense of interdependence is heightened to the nth degree as each of the partners feels that his own work gets the benefit of the best that the others have it in them to give. It is the FR principle that provides the basis for the emotional unity of the class-stratified Platonic polis.

My presentation of Plato's theory has been sympathetic. I have seen no need to regurgitate familiar critiques of it. Here, as in *SJR,* I have thought it more profitable to try to show what there is in this theory which, in spite of everything that can be said against it, is still, quite recognizably, a theory of *justice,* and this in spite of the fact that in his exposition of it Plato chooses to isolate it from the notion of equality. My argument in both essays has been that this occurs only on the surface—that, if one looks at the way the theory works, one will find that equality still does considerable work in it. In the earlier essay I stressed the work which formal equality, i.e., impartiality, does in shaping Plato's vision of a just polis. Certainly this is important, and quite sufficient in and of itself to undercut the most virulent of the critiques of Plato's theory of justice that has been published in my lifetime—Popper's.[38] However, I failed to indicate in that paper that while impartiality is indeed a necessary condition of justice, it falls so far short of being a sufficient condition that there are many circumstances in which its exercise may even confirm, instead of blocking, injustice. If the prevailing allocation of rights is itself unjustly unequal, those who act with impartial regard for those rights may be thereby only abetting and consolidating the injustice. White persons acting with scrupulous impartiality to respect the rights assigned to whites and blacks under the segregation laws of South Africa may make it only more certain thereby that blacks will be kept out of the jobs, schools, living space, etc., from which they are being excluded by those laws. So if Plato offers a theory of rights of persons within the polis which, however different from ours, is still a *bona fide* theory of justice, he must be putting into it something more than formal equality: he must be injecting a measure of equality into its content as well. This is what I have located in the FR principle, and have sought to show how this operates as a constraint on unjust inequalities, thereby producing a design for the ideally just polis that is noticeably closer to our own sense of justice than it would otherwise have been.

However, what the FR principle cannot provide is a basis for substantively equal "human" rights. It will justify equal rights only in those special cases in which the differences between groups of persons (such as differences of sex) are judged to be irrelevant to the value of their respective contributions. When these differences *are* relevant, rights cannot be equal: this is the inexorable implication of the FR principle. Inequalities in congenital capacity as vast as those between philosophers and producers thus entail correspondingly vast inequalities of right— inequalities so extreme that Plato finds it perfectly natural to represent the relation as idealized slavery. Here, if anywhere, one feels, something has gone decidedly wrong. What?

One way to go in looking for the answer would be to probe the metanormative foundations on which the moral theory rests— to examine the empirical and logical soundness of its psychological, metaphysical, and epistemological supports, all of which are so essential to it, that if even one of them cannot bear its due measure of structural weight the whole edifice will come tumbling down. In *SJR* I pointed out that Plato himself came to see, later on in his life, that two, at least, of the psychological assumptions were rotten and that this new insight doomed his earlier faith in the realizability of the ideally just polis.[39] In the *R.* Plato had believed that men deprived of every right to political freedom could coexist with their masters in affectionate solidarity. In the *Laws* he has come to see that this would be a mirage. He now confesses that "slaves and masters could never be friends" (757A). In the *R.* he had scorned the Herodotean insight that autocratic power corrupts the autocrat.[40] In the *Laws* he has rediscovered that truth. He now concedes that it is simply not within men's capacity to live in absolute command of the bodies and souls of their fellows without "becoming filled with *hybris* and injustice" (713C), and thus unfitted for the function which, on his own view, only incorruptibly good and wise men could justly discharge.[41] So if one were only interested in showing Plato that the theory that led him to deny human rights to the mass of the civic body was disastrously wrong, it would suffice to elaborate just these considerations. Any theory of justice which is predicated on an erroneous estimate of human capacity can be rejected on just that ground.

But such a line of refutation could only show that Plato's theory was unworkable. That it was also unacceptable on moral grounds—that its vision of ideal justice was itself flawed—it would not show. Can we convict it of this deeper failure without indulging in moral parochialism, faulting it simply for enunciating principles of justice that disagree with ours? I believe we can. We may appeal to the common morality of his own time and place. This matches ours—and indeed every other single-standard morality known to me—in being equalitarian through and through in one fundamental respect: its basic rules confer on all members of the community rights which are equal in a sense approaching the ideally full sense defined above (p. 182):

1. all have equal right to share in those benefits which accrue to individuals from the general observance of moral rules, and, moreover,

2. all have equal right to share in making the judgments of praise and blame which guide the conduct by which these benefits are produced.

To see the import of (1) consider the benefits we all derive from the observance of rules interdicting murder, theft, kidnapping, blackmail, rape, slander, insult, and the rest. To realize how enormous are these benefits to each of us we need only imagine what life would be like in a society where the rule, "Thou shalt not kill," was not in force, so that forebearance from murder had to be purchased individually from those around us. What price would we be prepared to meet if they were driving a hard bargain? Would one million dollars to each be too much if we had the funds? The same—crudely melodramatic, but still true—line of reflection, applied to each of the other rules, will bring home the immensity of the benefits we derive gratis from the general adherence to the elemental moral interdicts on certain lines of socially injurious conduct. In entitlement to these benefits all members of the community are on a par: one person's right not to be murdered, robbed, kidnapped, etc. is exactly the same as that of anyone else. The smart and the dull, the productive and the unproductive, the moral and the immoral, have exactly the same right to be shielded from such assaults on their welfare, freedom, and dignity. When Rashkolnikov excuses his murder of old Alëna with the thought that humanity would be none the worse for being rid of such a miserable specimen—no one would miss that human louse—he violates the spirit, as well as the letter, of the moral law.[42]

To see that (2) is true, one need only consider the fact that the individual judgments that guide the rule-observing conduct by which those benefits are produced consist in the main of that praise or censure which every person is entitled, equally with every other, to pass on his own personal conduct and on that of his fellows in respect to its conformity to the rules. This precious right is not reserved to a conclave of moral pundits. Though the judgment of persons known for their moral sensitiveness and sagacity may well be entitled to high respect, still, in the last analysis, each moral agent must make his own judgment, incorporating whatever insight he derives from others, but judging ultimately for himself.[43] No matter what may be the role of leaders credited with special moral insight on certain questions, the judgments that guide conduct in the community have to be made by the people themselves whose lives *are* the life of the community, the ones who act and are acted upon, and whose approvals and disapprovals shape their own actions and those of their fellows: morality, if it works at all, will only work in that way. To pass such judgments all persons are equally entitled: the right of Sonia, the prostitute, to give Rashkolnikov her own direct, spontaneous verdict on his crime is the same as that of anyone else.[44]

By confronting Plato with these two considerations we could show him that his adherence to the view that only functional rights exist puts him at odds with the morality of his own society, not merely with our own conceptions of what is just and right. Thus, in taking this view in respect of (1), he is implicitly denying that persons have certain basic rights conferred on them solely by the moral rules. And to deny this is to give up the commitment to the single-standard morality which confers those rights. Plato does this openly in the case of the interdict on lying. His acceptance of a double standard here is blatant. The duty that subjects have to refrain from lying when addressing their rulers is not reciprocal: rulers have the right to lie to their subjects whenever they judge this would be good for the state.[45] Moreover, in respect of (2), Plato's theory gives the elite a monopoly on authoritative moral judgment. Their subjects will, naturally, be making moral appraisals of each others' acts as they live their lives from day to day. But in so doing they will be either repeating the ones inculcated in them by the philosophers or, if they try to do more, will be only hazarding verdicts whose validity will be worthless until stamped "valid" by higher authority. Thus even within their own circle these people will be in no position to make authentically personal moral judgments; their conscience will be a shadow-conscience, of no account except insofar as it mimics or anticipates that of their rulers. It follows a fortiori that the populace has absolutely no right to pass moral judgment on the most important by far of all the actions taking place within their polis: those of their rulers, which may affect in minutest detail the life of every citizen. So the philosophers are beyond everyone's moral judgment except their own; morally they are a law unto themselves.

Should Plato be troubled by the fact that his theory clashes in this way with assumptions deeply entrenched in the morality of his own time and place? He should. For he knows his own dependence on that morality. Though he thinks its precepts "vulgar," he cannot ignore his debt to them.[46] He speaks of them with affectionate reverence: "they have been like parent to us; we have obeyed and revered them."[47] He is caught in a bind here. For he must know that the validity of those unreasoned precepts is presupposed by his own reasoned conception of the justice of the mutual benefit society. They are the indispensable, though unacknowledged, base on which his whole construction rests both at the first level of the protopolis and at the highest level of the fully developed ideal polis. At the former the dependence is only implicit, but nonetheless real: were it not for the general observance of the rules of common decency, the life of protopolis would quickly degenerate into a jungle of fraud and violence. The dependence comes close to becoming explicit at the higher level. Since he insists that the ideally just man of the ideally just polis is bound to observe the "vulgar" standards,[48] Plato must have some awareness of the fact that the practice of justice as defined by himself requires fidelity to the common precepts of morality. But he never confronts directly this all-important fact, and hence is never impelled to explore its limitations.

The failure may be traced back to the very method he employs to discover the rights of persons in the polis. The purely a priori method by which he constructs the "perfectly good" polis, abstracting systematically from all empirical data, including those constituted by the precepts of the prevailing morality, relieves him of the obligation to explain and justify the multitude of bonds which must exist between his new conception of justice and the one embedded in common opinion. Had he shouldered this task he might have seen that the base on which his whole construction rests already accords to persons in the polis a complex of substantively equal rights which are totally independent of differential ability to excel in socially productive roles. He might then have sensed the one-sidedness of his theory of justice. What it offers is an idealization of the justice of the work-ethic—the domain in which, generally speaking, persons must earn their rights through their productive labor. What it fails to recognize is that persons are also related to one another as persons, not merely as functioning units in a productive enterprise, and that as so related they have already as their birthright in their common morality substantively equal rights and may claim for the like reason substantively equal rights to participate in the governance of the polis.[49]

Postscript on the Use of "Rights"

I have not felt it necessary to introduce my use of this term in this essay by way of a moral analysis; its preanalytic use has served my purposes well enough. However, it may be well to indicate summarily my general understanding of this term, reserving its exposition and defense for another occasion. Accordingly, I submit the following contextual definition: A substitution-instance of the sentence form "A has the right to X against B" will be true for persons bound by a given moral or legal code if and only if B is required by the norms of that code to engage in X-supporting conduct (action or forbearance) demandable of B by A and/or others acting on A's behalf. The norms of Plato's ideally just polis would satisfy this condition. In so doing they would confer rights ipso facto, that is to say, without anything having to be said about rights in its code or in talk about it. If persons in the polis, or others acting on their behalf, are in a position to demand rule-mandated conduct beneficial to those persons, then they *have* rights under its code, regardless of the words at their disposal for naming or describing rights. I emphasize this point because there is no special word for "rights" in Plato's mother tongue—no word which corresponds exactly to ours, behaving as it does in all the contexts in which we speak of rights. To express the same notion one would resort to a variety of makeshifts: (1) "what is due to one" (*ta opheilomena*), as in the definition of "justice" ascribed to the poet, Simonides, in *R.* 331E, "rendering to everyone his due"; (2) "the just" (*ta dikaia*), as in Demosthenes 15 (*Rhodians*), 29, "in commonwealths the laws have made participation in private rights (*idiōn dikaiōn*) common and equal for the weak and the strong"; (3) "one's own" in the phrase "to have one's own" (*ta hautou echein*), as in the definition of "justice" in Aristotle's *Rhetoric* (1366B9-10), "the virtue because of which each has his own and in conformity with the law."

In an influential paper ("Are there Natural Rights?" *Philosophical Review* 64 [1955]: 179-91) H.L.A. Hart has maintained that the existence of rights is a contingent feature of a moral code. He writes:

> There may be moral codes quite properly termed moral codes . . .
> which do not employ the notion of a right, and there is nothing
> contradictory or absurd in a code or morality consisting wholly of
> prescriptions or in a code which prescribed only what should be
> done for the realization of happiness or some ideal of personal
> perfection. (p. 176).

I would maintain that, on the contrary, any moral or legal code necessarily confers rights if it satisfies the elementary condition I have specified above, which all historical codes known to me do satisfy, not only those of classical Greece but even those of the very earliest forms of social organization: see, e.g., E. Adamson Hoebel, *The Law of Primitive Man* (Cambridge, Mass.: Harvard University Press, 1954), a study of the legal-moral codes of conduct in a variety of "primitive" societies (several of them having prepastoral economies), in all of which the concept of rights is profusely instantiated and, indeed, proves the fundamental concept required for the description and analysis of the empirical data.

Ignoring empirical findings such as those presented in Hoebel's study, Hart adduces the Mosaic Decalogue as a historical example of his thesis. He claims that it cannot be interpreted as conferring rights because

> in such an interpretation obedience to the Ten Commandments
> would have to be conceived as due or owed to individuals, not
> merely to God, and disobedience not merely as wrong but as a
> *wrong to* (as well as harm to) individuals (p. 182).

Accepting the implied criterion for the existence of rights in a given code, I would expect it to be satisfied perfectly in the case of the Ten Commandments. Thus, consider someone who was cheated out of his patrimony by false witness. To concur with Hart we would have to believe that it would be possible for the *general response* to such an event (i.e., not the response of some obsessive fanatic, but that of the average Israelite) to include awareness of the crime as (a) sinful and wrongful disobedience to the code, while excluding awareness of it as (b) a *wrong to* the victim. If Hart has any evidence that the ancient Israelites were capable of making habitually the extraordinary abstraction of (a) from (b), he fails to allude to it. I know of none in the pages of the Old Testament or of any other record of human experience.

NOTES

* This paper is a companion piece to my earlier essay, "The Theory of Social Justice in the Polis in Plato's *Republic*" (to which I shall refer hereafter by the abbreviation *SJR*), in a symposium entitled *Interpretations of Plato* ed. Helen North, as a supplementary volume to the Dutch classical journal, *Mnemosyne*, 1977, pp. 1-40. For certain aspects of Plato's theory—particularly the foundations of the social theory in Plato's moral psychology and metaphysics, which I have not attempted to develop in the present paper, I must refer the reader to *SJR*, sec. 3, pp. 26-34.

1. In this respect Plato's theory of justice rests on a "historical" base in the technical sense of this term introduced by Robert Nozick, *Anarchy, State and Utopia* (New York: Basic Books, 1974), pp. 152 ff.: a historical contingency (in this case, the accident of civic birth) is to determine who are to have the rights defined by Plato's theory.

2. For slavery as an integral component of the economy of classical Greece see especially M.I. Finley, "Was Greek Civilization Based on Slave Labor?" *Historia* 7 (1959): 145-164.

3. *R.*, 433D. Other references to slavery are even more problematic. For my discussion of all the passages which have any bearing whatever on the question see my paper "Does Slavery Exist in Plato's *R.*?" *Classical Philosophy* 63 (1968): 291 ff., reprinted in my *Platonic Studies* (hereafter "PS") (Princeton: Princeton University Press, 1973), pp. 140 ff. The best argument to the contrary is in R.E. Levinson, *In Defense of Plato* (Cambridge: Harvard University Press, 1953), pp. 167-73.

4. Cf. the preceding note.

5. John Locke, *Second Treatise of Civil Government,* chap. 2, para. 14. These are rights which Locke takes to exist in "the state of nature" and cannot be overridden by the positive laws of any society at any time in any place.

6. Or, more literally, as "each making his own work common to all." (For the translations I shall be using see the list in n. 49 below.)

7. Karl Marx, *Capital,* ed. Friedrich Engels; translated by S. Moore and E. Aveling (New York: *Modern Library),* chap. 14, sec. 4, pp. 400-401 and notes.

8. There are some hired laborers (371C-D), but presumably only as aids to self-employed producers. For enterpreneurs as such no place whatever is made in this primitive economy.

9. R.C. Cross and A.D. Woozley, *Plato REPUBLIC: A Philosophical Commentary* (London: Macmillan, 1964), p. 80.

10. The first sense listed for "selfish" in the *Oxford English Dictionary* is "deficient in consideration of others," and this is surely by far the most common use of the word. In a community of "partners and helpers" (369C3), where everyone is willing and able to pull his weight, there would be no special reason why everyone should be "deficient in consideration of others": everyone would be in the happy position of seeing that others benefited as much from his labor as he did from theirs.

11. This is the strength of the Cross-Woozley interpretation. Italicizing the phrase that Plato uses in 379C7, "thinking it is better for himself to do so," they gloss:

> The attitude which Socrates ascribes to each of his imaginary citizens is that of putting into the common stock . . . whatever he must in order to get out of it what he wants. That another man might have need of what you produce would not be regarded by you as a reason for your supplying it, unless you yourself had need of what he produces *Plato REPUBLIC: A Philosophical Commentary,* p. 80).

This is perfectly true, but all it entails is that these people are not being described as altruists—from which it does not follow that they are being portrayed as egoists. What does follow is that, they, any one of them, *could* be an egoist, if so inclined, and get along quite well in that society—which is not to say that this is what all, or most, of them would have to be or what Plato expects them to be.

12. The Myth of Metals begins, "All you are brothers in the polis

. . ." (415A); the soldiers "must think of the other citizens as brothers . . ." (414E). And see the development of this theme in 461-464B, and the brief, but highly significant, reassertion of *philia* (in this case specifically between the intellectual and the manual worker) at 590D5-6.

13. Cross and Woozley, (*Plato REPUBLIC: A Philosophical Commentary*, pp. 91, 93, quote the latter phrase, admitting that it does not sit well with their interpretation. That Plato himself should feel no strain between that phrase and the one he had used at 369C7 gives some indication that the earlier characterization had not been meant to imply that the cooperative conduct was "entirely selfishly" motivated. Some indication of the same thing had also been given in 369C3 by the description of its people as "partners and helpers."

14. Cf. n. 2 above. For this use of *isos* see H. G. Liddell and R. Scott, *Greek-English Lexicon* (new ed., H. S. Jones, Oxford: Oxford University Press, 1925), *s.v.*, sense II, 1-3, and PS, 184-85, n. 78.

15. "Pareto optimality" would be preserved, and would be still, even if the inequalities were immeasurably intensified, as they might in given circumstances if the natural monopoly enabled its holder to perform a service of such importance to others that they would be prepared to give almost anything in exchange for it. Thus suppose that someone in protopolis had a unique medical skill, and his price for exercising it would be that the patient, on recovery, would give to him (or to one of his friends) a whole year's full-time personal service. If people were willing to accept these terms (their regular workload being shifted to friends of theirs willing to work overtime for their sake), Plato's stipulation that each would reckon the exchanges "better for himself" would be sustained and Pareto optimality would be preserved: judging the benefit to each by what he and his friends are prepared to pay in a legally unconstrained bargain, we would have to say that the community is better off paying this tribute to their medico than they would be by foregoing his service to them.

16. It does on mine, and I leave it to the reader to judge whether or not it does on his. I write in full view of a current conception of justice whose proponents could swallow without discomfort the example in the text above and, if they have strong stomachs, even the one in the preceding note. So, I assume, would Robert Nozick, if he adhered consistently to the views he expounds in his remarkable book, *Anarchy, State and Utopia*: if, as he holds, everyone has exclusive right to the exploitation of his own individual abilities, the heroes (or villains) of each of my two examples would violate

no rule of justice by acting as they do. Neither would the dis-
coverer of a cure for cancer whose formula, if made public, could
be mass-produced to cure thousands, but prefers to keep it secret
and its production scarce to enhance thereby the financial returns to
himself: cf. Nozick's own discussion of related (if less gruesome)
examples, on pp. 181-82 of his book.

17. *Ergon* ("work," "function"), and the verb *ergazesthai* (369E;
374A-C; 406C; 407A1 434A); *epitēdeuma* ("practice," "occupation," "pursuit"), and the verb *epitēdeuein* (374E, 394E, 359C,
433A, 454B-455B, 457B); occasionally also by *technē* ("art,"
"craft": 370B; 374A, C, E), which had been frequently inter-
changed with *ergon* in Book I). *Dēmiourgos* ("craftsman") is used
repeatedly of the rulers (395B, 421C, 500D) in defiance of the fact
that these people are rigorously excluded from the *manual* crafts for
which *dēmiourgos* and *dēmiourgia* were commonly used (and are so
used by Plato in the same Books of the *R.*: 371A, 406C, 468A).

18. For textual documentation see the passages adduced in support of a
similarly expanded description in PS, 119.

19. See n. 17 above.

20. Plato repeatedly uses the adverb *kalōs* ("beautifully") to refer to
excellence in the performance of social function, as e.g., at 370B,
374A, 394E.

21. Thus the philosopher-ruler is described as a "craftsman"
(*dēmiourgos*) who "moulds himself" (no less than the polis) in the
image of his ideal vision (500D).

22. I take equality and reciprocity as entirely distinct notions, and will
so use the terms in this paper. The warning is necessary since "reciprocity" is sometimes used in moral contexts so as to *entail*
equality: so, for example, in Jean Piaget's admirable study, *The
Moral Judgment of the Child*, trans. M. Gabain (New York: The
Free Press, 1965), where it occurs interchangeably with "equality"
and "equity"—understandably so, perhaps, since the only kind of
reciprocity Piaget considers relevant to the moral education of the
subjects of his study (children nurtured in a Western democracy
imbued with a liberal ethos) is *equalitarian* reciprocity within the
children's peer group in contrast to their submission to authoritarian
standards imposed on them by adults.

23. *SJR*, pp. 19, 22.

24. For references see *SJR*, p. 22, n. 79.

25. *R.*, 544C and ff.

26. And cf. his description of how oligarchy breeds within it the forces
which will destroy it and bring in democracy: 551C-556E.

27. That is to say, substantive, as distinct from merely formal equality. Cf. *SJR,* p. 20, n. 69.
28. Plato's feminism is ably defended in the following recent papers: Dorothea Wender, "Plato: Misogynist, Paedophile, and Feminist," *Arethusa* 6 (1973): 75 ff.; Christine Pierce, "Equality: *Republic* V," *Monist* 57 (1973): 1 ff.; W.W. Fortenbaugh, "On Plato's Feminism in *Republic* V," *Apeiron* 8 (1975): 1 ff.; Brian Calvert, "Plato and the Equality of Women," *Phoenix* 29 (1975): 231 ff. My own argument for the thesis will appear in a book on Platonic Justice that I hope to publish shortly. For the best argument on the other side see Julia Annas, "Plato, *Republic* V, and Feminism," *Philosophy* 51 (1976): 307 ff.
29. This becomes most strikingly clear in a passage at the concluson of Book VII, which seems to have escaped the notice of Annas in her otherwise well-documented paper mentioned in the preceding note: Socrates speaks of the guardians who "have survived the tests [for sagacity and virtue] and have excelled in all things in action and knowledge," have come to know the form of the Good, have taken their part in governing the polis, and at their death receivd quasi-divine honors from the polis. When Glaucon remarks, "you have made your men-rulers [*archontas,* the masculine form] surpassingly beautiful, as in a work of art," Socrates retorts:

> Yes, and the women-rulers [*archousas,* the feminine form] too, Glaucon, for you must not suppose that my description applies to men any more than it does to all the women who arise among them with the adequate natural endowment (540C3-6).

30. A point well brought out in the paper by Annas.
31. *R.,* 451D ff.
32. See e.g., G.R. Morrow, *Plato's Cretan City* (Princeton: Princeton University Press, 1960), the references to "women" in the general index to this book.
33. *R.,* 428A11-429A3. The basic qualification for the function of governing, "knowledge" or "wisdom," is announced at this point. The central books of the *R.* (475B to the end of Book VII) fill out the import of this vast claim. Cf. also *Politicus,* 292C5 ff. on "the most difficult and greatest" knowledge, the basis of the statesman's authority, which barely fifty out of a thousand citizens could hope to have—perhaps "only one or two or very few of these" (293A2-4).
34. Imputing to his philosophers his own boundless enthusiasm for

intellectual work, and his distaste for managerial office, Plato feels that this political work will be a grave sacrifice of personal happiness for the good of the commonwealth and argues earnestly that it is just to impose this painful duty on them (519C-521B).

35. *SJR,* pp. 28-30.
36. That this is the correct rendering of *homoioi* in this context I have explained in *PS* (p. 13, n. 34). Cornford's "equals" would be too strong as a translation, though one can sympathize with his reasons for so rendering it: equality is not too far from the word's ambience, as shown by the common use of *homoioi* for the "peers" at Sparta (all of them equally privileged by contrast with the legally underprivileged rural population, the *perioeci,* and the cruelly oppressed Helots), and by Plato's association of it with the leveling of social ranks which he thinks characterizes democracy (562D8 and E7.) Cf. also the use of *homoioi* in Thuc. 1, 34, 1.
37. And cf. the other references in n. 12 above.
38. Karl Popper, *The Open Society and its Enemies,* 4th ed. (New York: Harper Torchbook, 1962).
39. *SJR,* pp. 36-37 and Appendix C.
40. Herodotus puts this in the mouth of a Persian opponent of tyranny who indicts it by observing that "even the best of men placed in such a position of [absolute] political power would be changed from his wonted mind" (3, 80, 3) and would be unable to hold back from *hybris* against his subjects.
41. The same thought in 691C and 875A-D.
42. "But I only killed a louse, Sonia. A useless, nasty, harmful louse" Fyodor Dostoyevski, *Crime and Punishment,* trans. David Magarshack (London: Penguin Books, 1966), p. 84.
43. This is the sense in which moral autonomy is absolutely indispensable to the morality of an adult person. To be denied this right is to be denied full moral capacity, which includes what Gerald Dworkin calls the right to "critical reflection"—the right to an independent examination of any expert ("Moral Autonomy," para. 47.)
44. She gives it without "critical reflection" and in less than a sentence in her instant response to Rashkolnikov's remark in n. 42 above: "A human being—a louse?"
45. *R.,* 389B-D; 414B-C.
46. *R.,* 442E.
47. *R.,* 538C.
48. *R.,* 442E-443B.
49. Translations of the *Republic* which I have used above:

Bloom, A. (New York, 1968)
Cornford, F.M. (New York, 1945)
Grube, G. (Indianapolis, 1974)
Lindsay, A.D. (London, 1935; originally published in 1906)
Shorey, P. (London and Cambridge, Mass.; Vol. I, 1930; Vol. II, 1935)

Commentary

Moral Autonomy and the Polis: Response to Gerald Dworkin and Gregory Vlastos

H. Tristram Engelhardt, Jr.

THERE IS A SIMILARITY, even if masked, between the arguments of Professor Vlastos and Professor Dworkin. Their central concern is no less than with the sociality of man. They recognize that we are embedded in a social context and that society shapes and gives purpose to our lives. In particular, a social context gives content to our moral lives. Both Dworkin and Vlastos address this point, though in different ways. They fail, however, to say much about the role of history and biology in shaping that content. Yet that content takes form through the nuanced constraints of history and biology. Any detailed account of social roles and rights requires detailed presuppositions about the capacities of humans. Any concrete treatment of the foundations of ethics will require recourse to science. Professors Dworkin and Vlastos at best hint in this direction—a suggestion I will explore at the conclusion of this paper.

In advancing an account of the principle of "functional reciprocity," which Plato used to justify a structure for the polis, Vlastos offers a notion of moral integrity. For Vlastos's Plato, the basis for justice is function in and for the polis. That is, equality and fairness are to be understood in terms of function within the polis. Dworkin likewise argues against a doggedly individualistic morality. He does this by suggesting that we rely upon those authorities whose moral competence has been determined (i.e., by weak checking, cf. Dworkin p. 168). We can accept the guidance of moral authorities, so Dworkin implies, without assessing each bit of advice they proffer. (Though Dworkin provides an explicit critique of "moral autonomy," he only suggests in conclusion that a "more adequate understanding" of "tradition, authority, commitment, and loyalty," could provide a conception of autonomy "worthy of admiration." See Dworkin p. 170. What his position would be can only be surmised from his criticism of moral autonomy.) Both Vlastos's Plato and Dworkin have little sympathy for accounts that portray morality as simply an individual matter.

Vlastos's Plato portrays moral practices as creating the status of individual moral agents. The status of moral agents is dependent upon the character of the moral practices that form their social ambient. The ideal case of that ambient is, according to Plato, the Republic—a moral environment that defines its component parts, its citizens. In contrast, Dworkin's moral agents define themselves with the help of the moral authorities they accept. But they, the agents, still bring critical reflection to bear on moral authority and tradition. Dworkin's moral agents maintain an explicit autonomy not given to the members of the Republic. Dworkin, in short, would appear to remain a defender of moral autonomy and pluralism. He continues to show interest in the integrity and conscientiousness of the moral agent and seems to aver approval of that moral autonomy exercised in "weak checking" (i.e., in justifying one's particular moral authorities), rather than in "strong checking" (i.e., in justifying all the principles for each moral choice). If this is so, then Dworkin is a rule autonomist rather than an act autonomist. Vlastos's Plato, however, gives a political account of rights with a focus on function in and for the polis, and without providing a mechanism for the members of the polis to engage even in the weak checking of Dworkin. In what follows, I will examine and compare their accounts.

The Ambiguities of Autonomy

I will begin with Professor Dworkin's paper. Dworkin contrasts two poles of moral discourse: (1) the autonomy of the individual moral agent (here autonomy is understood to be a form of individualistic authenticity), and (2) the objectivity of a moral order that would allow one to judge the status of moral claims. In other words, the moral autonomy of the individual moral agent (i.e., the state of having formed moral judgments in a fashion true to oneself) is contrasted with the objectivity required for the moral life (i.e., required to insure the integrity of social institutions and of the social enterprise). Dworkin emphasizes that decision making in the ethical sphere requires a sense of moral objectivity so that one can assign significance to individual moral choices. The difficulty is that while proceeding to erode the legitimacy of the claims of moral autonomy as authentic moral individualism, his background use of morality continues to rely heavily on the notion of autonomy. In addition Dworkin gives very little account of how traditions, institutions, and authorities provide the moral roles and contexts of the moral life. To supplement this deficiency he would need to provide a counter-balance to individualistic autonomy somewhat in the vein of Vlastos's Plato. In fact, Dworkin focuses so narrowly on the failure of individualistic autonomy that he fails to do much more than to allude to the importance of traditions in providing the institutions and roles that give the moral life completeness and actual purpose. As a result, the objective order remains a structure that is potentially alien to individual agents, rather than the source of their concrete moral lives.

This tension between the morality of the abstract autonomous individual and that of the concrete social individual is indicated in Dworkin's epigraphic quotations from Kant and Hegel. The quotations are both helpful and misleading. They are helpful in suggesting the dialectical move that Dworkin wishes to make from an understanding of virtue as individual integrity realized in and through social conduct. The quotation from Hegel is thus useful in indicating a direction for the considerations at hand. The quotation from Kant, though, is out of place, for it suggests that Kant (despite an appropriate qualification in the text of the paper—see section 23) construed moral virtue as autonomy in individualistic moral choice. It is true that Kant begins the *Foundations of the Metaphysics of Morals* by asserting that "nothing in the world . . . can possibly be conceived which could be called good without qualification except a *good will*."[1] But the good will, which for Kant is the autonomous will, is anonymous in its moral activities. It is precisely a good will when it wills *not* in terms of the peculiar or particular inclinations of the moral agent involved, but from the general viewpoint of a rational free agent, of a member of a world of rational beings, a member of the *mundus intelligibilis*.[2] The Kantian moral agent does not worry about satisfying conditions (1) and (5) of Dworkin. Moreover, he or she need not, in fact should not, worry about satisfying any of the conditions (1) through (5), if the accent in those conditions is placed upon "his" as designating moral principles *for him* or as binding *upon him* in his particularity. Kant calls upon us to view ourselves as judging anonymously when judging morally, that is, judging without particular consideration of ourselves—"Act by a maxim which involves its own universal validity for every rational being."[3] The authenticity of the moral agent is thus not the authenticity of *that* moral agent but of *any* moral agent.

Where, then, is the bite of Dworkin's attack on autonomy, beyond impeaching those who would hold that autonomous actions are idiosyncratic actions? It would not seem to strike home to Kant's notion. Kantian construals of moral autonomy have an intrinsically nonindividualistic character which makes Dworkin's criticism of them irrelevant. Kant is advancing, albeit in a one-sided fashion, a picture of the moral agent as essentially social—in all choices the autonomous moral agent is the moral agent who chooses as any moral agent should. He or she chooses for all. Every moral agent is in this sense always a member of the community of all moral agents, considered anonymously. Moreover, to choose autonomously is, for Kant, precisely to be determined as an individual according to the universal characteristics of moral agents. "Therefore a free will and a will under moral laws are identical."[4] In this way, Kant attempts to integrate (1) the moral agent as an individual, and (2) the objectivity of the moral order. He embeds the individual's autonomy in the moral order. Individual autonomy becomes equivalent to acting out of and within the moral order.

But Dworkin fails, as does Kant, to distinguish consistently between autonomy as a side constraint (i.e., self-determination, in the sense of free choice, as a necessary condition for the possibility of being worthy of blame or praise),[5] and autonomy as a particular value to pursue (i.e., being self-determining in the sense of being influenced by others, by passions, etc.). These two senses of autonomy are conflated, for example, in Kant's arguments against suicide and masturbation.[6] But if one imposes such a distinction, one can then consistently speak of someone autonomously deciding not to be autonomous (e.g., deciding freely to get drunk, or to accept the authority of another). Kant, as well, welds together the concepts of choosing according to the moral law (i.e., autonomously in the sense of determining one's actions qua moral agent under the general formal constraints of the moral law) and being imputable for one's actions (i.e., responsible for one's particular moral actions). The result is a richly ambiguous concept of the particular moral agent acting as a moral agent in general (and therefore outside of a particular history, tradition, moral institution, or social role). The claim that moral agents ought to act autonomously means for Kant that moral agents are self-determining, can be held responsible for their actions, and act autonomously *in sensu stricto* when they act as moral agents should in general. The validity of moral claims does not, according to Kant, depend upon their being accepted by a particular moral agent.

It is by the strategic ambiguity of this concept that Kant escapes the paradox Dworkin raises in his section 21. There Dworkin wonders how one could rest morality on the agents' will and still claim any objectivity for the maxim, "moral agents ought to be autonomous." Would not saying that one ought to be autonomous ground that ought elsewhere than on the agents' will, i.e., heteronomously, nonautonomously? If the morality is grounded on the individual will, it is autonomous but nonobjective; if it is objective, it is not grounded on the individual will and is therefore not a morality of autonomy. But Kantian morality, though abstract, can still consistently assert that one ought, as an autonomous, individual moral agent, will in accord with the nature of one's will as a moral agent, i.e., rationally and universally. Dworkin does show that one cannot ground morality on individualistic decisions of individuals, *simpliciter*. Yet, in that

Kant distinguishes between the capacity to choose freely (*Willkür*) and the goal of choosing as a rational free agent (i.e., according to the moral law (*Wille*),[7] he is able to assert that moral agents ought to be autonomous, ground that claim on the nature of the agent's will, and still be coherent (even if, perhaps, not too helpful). Autonomy for Kant is not acting in terms of the idiosyncratic will but according to (one sense of autonomy) the nature of the will in general (*Wille*), while being (a second sense of autonomy) accountable for all one's decisions (*Willkür*). The paradox which Dworkin alleges, insofar as it exists, would require a better specification of one of the multiple meanings of "acting autonomously."

What is also needed is an account of the concrete social moral agent embedded in history, moral institutions, and social roles. (Dworkin does indicate that traditions are important, but usually through an argument in favor of objective moral reasoning, and without a developed corollary argument on behalf of the concrete particularity of moral institutions). Hegel's critique of Kant, as Dworkin cryptically suggests, is an account of moral choice that rejects a simple morality of objectivity through universalizing; Kant's position is rejected by Hegel because it is too abstract to be of any use for particular moral situations. Sartre's troubled young man, trying to decide between joining the Free French or staying with his aged mother, is "obliged to invent the law for himself" precisely because it is not all clear how one usefully universalizes in such circumstances, if one is choosing as an ahistorical, individualistic moral agent. One can universalize convincingly in any direction. As Hegel observed, the infinite autonomy of Kant is merely "an empty formalism" (because it lacks a reference to the context of a culture, an ethos, a *Sittlichkeit*). (See G.W.F. Hegel, *The Philosophy of Right,* #135) In contrast, moral imperatives acquire content in a social ambient, in a civil society, and finally within a state, within which priorities are ranked. This appears to be part of Dworkin's point when he defends the need for appeals to moral authorities (the *Sitten* of Hegel's *Sittlichkeit*). Autonomy as moral individualism is, as Dworkin suggests, incompatible with objectivity and morals. It is anarchy, the opposite of the well-integrated polis or civil society.

It cannot account for obligation. Autonomy in the sense of moral individualism, and autonomy in the sense of an abstract universalism, both fail—though each for different reasons.

Before Autonomy—Living for the Polis

We are brought, then, to Professor Vlastos's thorough and very useful analysis of Plato's sketch of civil society. In answer to the question of how Plato ascertains the rights of members of the polis, Vlastos refers to a principle of functional reciprocity. The move from the protopolis to the polis (*Republic* 2.369 sq) involves a move in conceptual richness from a barter economy in which there is individual self-sufficiency, to an economy where individuals play roles within an enterprise which has a life of its own. The move is from a stage of economic individual self-sufficiency to a matrix of economic interdependencies in which one does best by "performing that single function pertaining to the polis for which his own nature is best fitted." (*Republic,* 4.433a 5-6) One finds one's significance in and through his or her role in the polis.

One may here be reminded of some of Hegel's exaltations concerning civil society as "the mediation of need and of one man's satisfaction through his work and the satisfaction of the needs of all others. . . ." (*Philosophy of Right*, #188). One accrues significance in being a member of a civil society. But in the Republic one does not, as one does in Hegel's state, maintain a moment of individual rights or interests. Thus, Plato's carpenter with a crippling chronic disease would prefer death to medical treatment that would support his life without allowing participation in, and contribution to, the life of the Republic. (*Republic* 3.407d-e) The individual realizes his or her good in a social context, which good is, in fact, a social good.

But why does Plato freeze out talk of equality from his considerations of justice, as Vlastos asks? Certainly not simply because the term was "too much of a political hot potato" (see Vlastos, p 181). Equality smacks of the autonomy and individualism that Dworkin criticizes. What is important for Plato is not individualistic autonomy or equality, but function and contribution

within the social fabric of the polis. Those differences of privilege which reflect real differences of function within the polis
are proper differences. Moreover, such privileges are not to be
viewed as private perquisities of atomic individuals who in aggregate form the polis. The principle of functional reciprocity, in
contrast with the principle of proportional equality, is a fundamentally social principle—its accent is upon the good of the
polis, not the good of a class. Plato is not primarily interested in
equality of persons or impartiality in the treatment of persons.
Rather,he is interested first in the character of institutions, and
the goods and social roles they make possible—including the
virtuous lives of individuals. He is not interested in individuals as
particular individuals, but as individuals capable of achieving
virtues, in fact, social virtues. Moreover, the standing of persons,
of the moral self, is derived from the character of the social
enterprise. Such an enterprise is a just political enterprise when
its members can live fully for its accomplishments. The just or
fair polis is the polis which assures the opportunity to contribute
according to one's ability. Even to speak of equality tends to
obscure the message by an account of the particularity of individuals, prescinded from, and prior to, their social function. The
focus of the discussion is the social enterprise, not the individuals
who institute it.

It is instructive to note how far the Republic is from having,
for example, a Rawlsian difference principle so that only that
disparity in distribution of resources is justified which maximizes
the good of the least well-off citizen of the polis. Such a sense of
fairness is not at work in the Republic unless one defines the
good of individuals not in terms of Rawlsian primary goods, but
in terms of contributions to the polis. Also, the members of the
Republic do not justify their standing in the Republic in terms of
a reference to a hypothetical contractual situation in which absolute equality is relinquished given certain considerations. They do
not create the Republic. Rather, the Republic creates them.

There are thus good internal reasons for not importing rights
talk into Plato's account insofar as rights language identifies
justifications for demanding forms of conduct from others (actions or forebearances) because of the kind of entity one is (e.g.,
a person, a moral agent). There are no basic personal rights for

the members of Plato's Republic. One can only make demands upon the conduct of others in Plato's Republic in virtue of what the Republic is. Plato is one-sided in his account of morality as is Dworkin's autonomous individual. An adequate account of rights and moral conduct would need to address (1) the limits placed upon the authority of those who govern the polis in virtue of the polis being composed of entities capable of moral responsibility (i.e., because of the respect due to persons); (2) the incoherence of a morality that treats individuals simply as atomic social agents (i.e., autonomous individuals); and (3) the abstract vacuity of viewing moral agents only as universalizing agents. But neither Dworkin nor Vlastos shows how these considerations are to be balanced. Vlastos, when confronted by the nonautonomy of the members of the polis, embraces Dworkin's principle of critical reflection (in fact, with strong checking; Vlastos p. 190-91 and 200). Yet, Dworkin embraces this principle only somewhat half-heartedly—he accepts it in the absence of any arguments to the contrary (Dworkin p. 170). In addition, for Dworkin, the choice between individualistic autonomy and social embeddedness appears to be a choice between nearly exclusive alternatives.

> We are born in a given environment with a given set of biological endowments. We mature more slowly than other animals and are deeply influenced by parents, siblings, peers, culture, class, climate, schools, accidents, genes, and the accumulated history of the species. It makes no more sense to suppose we invent the moral law for ourselves than to suppose that we invent the language we speak for ourselves.
>
> Dworkin, p. 158.

Of course we *do* in part fashion and invent our language—we can become critical, reflective speakers—perhaps as Vlastos suggests, critical bearers and shapers of our social (and we might add, biological) destinies.

The Historical and Biological Constraints: A Critical Symbiosis

To succeed in such critical reflection one must be clear about the factual constraints that form one's theater of action. The importance of such empirical information is clear in Vlastos's account of Plato. Plato not only ignored the status of individuals apart from their role in the polis, he also assumed falsely that the nonintellectual man is naturally depraved, and the intellectual man capable of incorruptible virtue. Plato's theory of justice is ill-founded, as Vlastos indicates, insofar as Plato bases the principle of functional reciprocity on such false assumptions requiring marked differences between social classes. It is ill-founded for empirical reasons; detailed philosophical social theory is dependent upon information forthcoming from the sciences. At the least, such information can suggest the reasonableness of certain assumptions—for example, regarding what will be to the benefit of the least well-off individual. Or, to put it another way, there will be a reflective equilibrium between our a priori judgments about morality and the facts of the world, which equilibrium will define our concrete sense of justice at any one time. On the one hand, Kant's moral agents are, qua moral agents, without a history, a culture, or the particular inclinations of humans; and the ahistorical, autonomous individualists whom Dworkin criticizes do not even have the community of the *mundus intelligibilis* within which to find moral structure. On the other hand, historically rich accounts such as those of either Plato or Hegel vex us with the intrusion of particular presuppositions concerning human nature and history, presuppositions we may not share. We sail between the Scylla of vacuous abstraction and the Charybdis of historical parochialism.

Confronted with such perilous sailing, we are forced to give consideration to endeavors such as Wilson's sociobiology, and the work of social psychologists and economists interested in helping us secure a more complete presentation of the lineaments of morality. Such individuals, in fact, seem anxious to give advice. Wilson, for example, states that "while few will disagree that justice as fairness is an ideal state for disembodied spirits, the conception is in no way explanatory or predictive with reference to human beings."[8] He suggests an age-specific morality, holding that individual human inclinations to patterns of behavior change with different periods in the human life cycle. He argues that there may, in fact, be a genetic basis for an ethical pluralism in the sense of the existence of predispositions to divergent lifestyles, and that the existence of a range of moral dispositions may have survival value.

Even granting such explanatory and predictive claims, it is not clear how one should take account of such information in appraising either the status of autonomy or that of rights within the polis beyond indicating (1) within what constraints free individuals are likely to choose freely, or (2) how they are likely to develop patterns of mutual constraint. One should also remember that there has been a traditional recognition of the biological basis of some age-specific moral problems (compare the way sexuality structures the morality of a nine-year-old as opposed to that of a nineteen-year-old). Moreover, though it is the case that states of affairs, including genetic characteristics, predispose us, *grosso modo,* to certain value judgments, there is no solid evidence for a genetic basis for the differences among humans regarding such judgments (except for conditions of a rather trivial nature, e.g., color-blindness, taste-blindness, and other rather narrowly focused differences). Even before such evidence is forthcoming, it is already clear that the ethical life is lived in a historical and cultural ambient, a life-world structured by the results of many individual and common evaluations. Yet, prior to any such ambient, there are general considerations with regard to the meaning of moral conduct which turn on the conceptual conditions for justified blame and praise, respect for persons—a point that Vlastos makes. This last set of considerations involves the abstract conditions of individual freedom and respect for the autonomy of individuals such as Kant which, as both Hegel and Dworkin indicate, become real only within the fabric of a particular culture and history.

We see, then, that the accounts of Dworkin and Vlastos are both one-sided, but in different ways. For Dworkin to complete a critique of autonomy, he must develop more fully what he takes to be the balance among (1) the general moral obligations to act autonomously in the sense of acting from the moral law; (2) the general empirical constraints on achieving any morality or moral society; (3) the special value which he forwards through criticizing autonomous moral individualism (i.e., the value of having a role in a society); and (4) the proper regard for freedom as the condition of worthiness for blame or praise. Vlastos must help us to understand how Plato supplies what Dworkin's autonomous individual lacks—a moral tradition and a society. He should also suggest how the polis would look given critical reflection upon authority by its members. Is the polis still possible under such circumstances? Do we either have the anomie of individualism or the totalitarian constraints of the polis? One needs an integrated critique of moral individualism and its "aggregate" morality on the one hand (i.e., that social morality which is the sum of human behaviors without any commitment to the life of the community), and of the polis as the sole origin of the status of the moral individual on the other. What is needed is something like a Hegelian dialect which affirms the importance of freedom, while indicating that such freedom is by itself vacuous without a moral ambient, a society. Such a dialectic must show how the moral society presupposes freedom of the individual in order to ensure that its social concerns remain moral concerns (e.g., the state as the domain of true freedom as presented by G.W.F. Hegel, *Philosophy of Right,* #260). But more than that, an account must be given of the basic constraints upon realizing both individual autonomy and the polis.

It is here that empirical issues, such as those raised by the paper of Professor Alexander in this volume, are important for understanding the concrete nature of the moral ambient of human moral agents. It is here as well that we are likely to learn something of the kinds of social institutions that can realize particular human moral goods. If it is acknowledged (1) that the state is a community of persons who can be held responsible for their conduct and which thus presupposes the moral autonomy of persons, e.g., the right and the duty to engage in critical reflection on their actions, and (2) that critical reflection (i.e., moral autonomy) is vacuous without a history, a culture, and a biology, *then* we can explore, as ingredient to understanding the foundations of ethics, the ways in which those different histories, cultures, and biologies enable different balances of autonomy versus the power of the polis. We should then explore how and to what extent the constraints of history, culture, and biology are plastic. The concrete foundations of ethics would be recognized as influenced by science and technology. Knowledge about ourselves, and our ability to act upon that knowledge through technology, will determine the full sense of our autonomy and the nature of our polis.

NOTES

1. Immanuel Kant, *Foundations of the Metaphysics of Morals*, trans. Lewis W. Beck (New York: The Bobbs-Merrill Company, 1976), p. 9; *Kants gesammelte Schriften*, Akademie Textausgabe, IV, p. 391.
2. Ibid., p. 57, IV p. 438.
3. Ibid., p. 56, IV p. 437-38.
4. Ibid., p. 65, IV p. 447.
5. Robert Nozick, *Anarchy, State and Utopia* (New York: Basic Books, 1974), pp. 28-35.
6. For Kant's position regarding suicide see, for example, *Metaphysik der Sitten*, Akademie Textausgabe (Berlin: Walter de Gruyter, 1968), VI, p. 422-24; for his opinions concerning masturbation see, ibid., p. 424-26.
7. John R. Silber, "The Ethical Significance of Kant's *Religion*," an introduction to *Immanuel Kant, Religion within the Limits of Reason Alone* (New York: Harper and Row, 1960) p. ciii-cvi.
8. Edward O. Wilson, *Sociobiology, the New Synthesis* (Cambridge, Mass.: Belknap Press of Harvard University Press, 1975), p. 562.

Self-Conflict in Ethical Decisions

Eric Cassell

THE NOTION OF PERSON is fundamental to any consideration of medicine as a moral profession, concerned with the welfare of individuals. It is not sufficient, however, to require physicians to treat the patient as a person without providing a wider understanding of the concept. One important aspect of being a person is that a person is someone who valuates—who makes decisions based on alternative values. In an earlier discussion of valuational thought,[1] I believed that I could move forward to see how that mode of thinking is used to make value decisions and the relationship of the thought to action. I was stopped by the realization that the same individual could make contradictory moral or value decisions, depending on circumstances, and yet the decisions, even though conflicting, would seem authentic to the person. This essay is an attempt to describe conflicting ethical decisions made by the same person to see how conflict arises and how it is resolved.

Let me begin with the cases that caused me difficulty and that seemed to require an explanation.

In 1971, I took care of a woman, Dora S.,[2] who was dying of an inoperable cancer of the esophagus. The diagnosis was confirmed by biopsy and, prior to my seeing her, she had received adequate radiation therapy. Cancer chemotherapy available at that time offered no real hope of improvement. The decision to

215

maintain her comfortably rather than just keep her alive was made in concert with the family and indirectly in conversation with the patient. She developed pneumococcal pneumonia, which was not treated, and died of the infection. House officers connected with the case vigorously opposed the decisions. They felt that every available treatment should have been tried, that no patient should simply be "allowed to die." Their position was common in 1971. In 1977 I took care of Esther R., who was developing generalized paralysis from amyotrophic lateral sclerosis. As the paralysis spread, she experienced difficulty in breathing. Although she knew her life would be short (perhaps months), she wanted a belt-type respirator that would keep her comfortable and allow her to remain at home until she died. Her sons agreed with the decision. Because her respiration was dangerously impaired, she was admitted to the hospital for a few days until her belt respirator arrived. The house officers objected to putting her on a mouth respirator for fear that she could not be weaned off. Despite assurances that the belt respirator would allow her to go home, they were against "just keeping her alive." She died during the first night of hospitalization, apparently because no respirator assistance was provided. Similar cases are now becoming common. Dr. Mark Siegler tells of a ninety-year-old man, otherwise well, who was admitted to his hospital with pneumococcal pneumonia. The house staff did not treat him because they did not believe patients should be merely "kept alive." After all, they said, he *is* ninety years old.

The decision of the house officers in the 1971 case and the 1977 cases were the exact opposite of one another. Further, both examples, especially the woman with terminal cancer of the esophagus, but increasingly, cases like Esther R., are typical of their respective times—1971 and 1977. In both Dora S. and Esther R., the individual nature of the patient did not seem to enter the decision. Dora S. could not be helped to maintain a meaningful life and the family agreed that she should be allowed to die without further cancer treatment—a position with which the patient apparently concurred. Esther R. and her sons knew her situation and wanted the belt respirator.

What accounts for these paradoxical decisions whose only major difference was the period in which they were made? I am going to suggest that in each instance a *physician self* made the decision. Why suggest an entity such as *physician self,* rather than merely saying the young physicians made the decisions? Using physician *self* rather than young physician suggests that another or at least a different self might have existed in each of the young physicians. It seems possible that if, in each case, the physician had been not only a doctor but also the child of the patient, he or she would have made a different decision—a decision similar to that made by the actual children of the patients. Two alternative possibilities present themselves. The first possibility is that as the patient's child and also as a physician, more information would have been available and, perhaps, have altered the decision. The other possibility is that as the patient's child, despite also being a physician, a different self would have been presented with or have made the decision.

If it is simply the case that as the child who is also a physician, more information would have been available, then it is true that in the actual instances these young physicians could have sought more information. They are trained to know that information about the person of the patient is important to medical decisions. The families were available, the patients were of approximately the same social background as the physicians and there was no language barrier—all factors that might otherwise have prevented their knowledge of the patients. Furthermore, there was so much discussion about the cases that the young physicians knew the wishes of the actual children and even those of the patients. Thus, the information was available to them. But on what basis can I entertain the possibility that if they had been the children as well as physicians, they might have decided differently? In other words, if one is the child of a dying patient, might one decide differently than if one is the physician of a dying patient?

Another case sheds some light on that possibility. A middle-aged female writer was traveling a far distance to get to the bedside of her aged and dying mother who had just been transferred from a nursing home to a hospital. En route, the daughter had firmly resolved that she did not want the doctors to do anything to prolong her mother's process of dying. When the daughter arrived, the doctor suggested that her mother was indeed dying but he believed that because the mother was bleeding she might be helped if she was given blood transfusions. The mother's situation was essentially as the daughter had believed (the mother was dying), and precisely as discussed with herself en route. Nevertheless, at the bedside, after talking to the doctor, she found herself in conflict and sought outside help in making a decision. It would be common to say that while traveling she was using "reason" and "logic," but that at the bedside "emotion" prevailed. But, as she said herself, and as our own experience of the world testifies, she was also filled with emotion during the trip.

If we add this case to the previous situations, the evidence suggests that in the instances of the young physicians (and the adult intellectual en route), a physician "self" (or an adult-intellectual "self") made the decision. In other words, it was not that the young physicians (or the adult intellectual en route) were lacking information when they made a decision contrary to the wishes of family and patient (or daughter-at-bedside). Rather, it seems that the decisions were made by a "self" particular (discrete) to the entity, young physician. A particular "self" in this context should be read as a "self" separately identifiable in action and across time but inseparable from the whole person.

Before explaining further what I mean by "self," I should explain what kind of definition I will try to give to the word. Stephen Toulmin, in discussing the kinds of knowledge we can have about the self (and about self-knowledge), makes clear that the concept cannot, usefully, be seen as a hypothetical-explanatory one "to help us explain newly observed phenomena outside the normal range of our psychological experience." Rather, the concept as used here is descriptive and meant to "mark new complexities and relationships within psychological phenomena that have, in less exact and detailed terms, been long familiar."[3] I am merely trying to clarify a behavior that seems, on reflection, to be quite common—apparently different and conflicting judgments made by the same person about similar issues (in the examples it is the care of the dying) and in trying to clarify the behavior, I hope to enlarge the conception of "self," its maintenance, and integration.

An earlier way of seeing the self comes from George Herbert

Mead: "The Self has a sort of structure that arises in social conduct that is entirely distinguishable from the subjective experience that forms it." "The Self . . . arises where the conversation of gestures is taken over into the conduct of the individual form." The awareness of the Self (self-consciousness) "is not simply because one is in a social group and affected by others, and affects them, but because . . . his own experience as a self is one which he takes over from his actions upon others. He becomes a self in so far as he can take the attitude of another and act towards himself as others act."[4] In these citations, Mead makes clear that the self that he describes is cohesive and internally consistent, not something called up for the moment or immediately changeable. He goes on to say that "The essence of Self . . . is cognitive: It lies in the internalized conversation of gestures which constitutes thinking, or in the terms of which thought or reflection proceeds. And hence, the true origin and foundation of self, like those of thinking, are social."[5] This is, essentially, self as a learned role.

We can extend the word cognitive. As Toulmin points out, it would be "highly artificial to treat the 'understandings' existing between different human agents as something restricted to the cognitive sphere to which affective and volitional components happen to be annexed quite accidentally."[6] Except for specific or particular uses, it would be well to drop those distinctions as not being generally useful. As I leave my car and proceed down the block, my hand tells me that I forgot to lock the car door. When I examine a patient, my hand tells me that the lump in the breast is larger than the last time I felt it. Following previous distinctions between (say) cognitive or affective, I could not begin to know in which category to place what my hand told me. It does not injure Mead's insight to include with the cognitive, affective, and volitional functions, as well as what the hand says to its owner and to others.

Remember that my inquiry is prompted because the original cases suggest that a person can have more than one "self" and that these "selves" can make different ethical decisions in the same setting. The ethical decisions can be in conflict and, therefore, the "selves" can be in conflict. The problems to be addressed are a better understanding of what I mean (and do not mean) by the entity "self"; how conflicting "selves" could be maintained within the same (normal) person, i.e., what maintains the borders and how adjudication between conflicting "selves" takes place and ultimately how, from such adjudication, the individual might grow more mature.

As a first step, I accept that the word "self" is a poor word because it already carries too much baggage, is too fuzzy, and means too many different things to different people. But I think the word is troubling precisely because some of the matters being considered here are unsettled within us as individuals and in the use of the term. The notion of "self" is fundamental to ethical concepts such as autonomy where a unity of self is implied despite the fact that in the exercise of our own autonomy we are often "of two minds." Or, in attempting to clarify autonomy of belief, Gerald Dworkin has difficulty finding that which is uniquely from the self and only from the self because there is no "self" that can be seen entirely separate from other persons.

As noted earlier, a "self" is an aspect of person separately identifiable in action but inseparable from the whole person. The actions (that make it separately identifiable) are in response to things, other persons, or their actions. The "self" is, in other words, relational. Since it is relational, the self is inseparable from language. One could arrive at the conclusion that the self is relational from the opposite direction—starting from language. It is possible to stand in relation to a person or an object, or to act or decide to act without language, but it is inconceivable that the relation, decision, or action will not be described or justified by language at that time or later. As Roberto Unger points out, "consciousness displays a peculiar paradox that poses the preliminary issue with which a theory of mind must deal. Consciousness implies autonomous identity, the experience of division from other objects and from other [persons'] selves. But the medium through which consciousness expresses itself is made up of the symbols of culture and these . . . are irreducibly social. When you speak of language or make a gesture, you perceive and communicate meaning in categories that are the common patrimony of many men. By what power can you and they speak to one another? It must be possible for each to view the other's statements and acts as signs of certain intentions. These intentions can, in turn, be understood, because they are intentions you too might have. It follows that consciousness always presupposes the possibility of viewing other persons as selves that could, under favorable enough circumstances, see what one sees and believe what one believes."[7]

The properties of (spoken) language allow individuals to place themselves in relation to concepts, things, persons, or their actions by the way words and syntax are used. Verbs, adverbs, adjectives, and pronouns are used to create what might be called a semantic space in which the self as speaker is central and the object of conversation is at a valuative, affective, or attitudinal distance from the self. A different relationship is described by "my mother is dying" and "my dying mother." Or, "my mother is suffocating to death," "my mother is dying because she can't breathe anymore," and "the woman is dying from respiratory paralysis." Although the sentences in both sets describe the same phenomenon, the distance between the speaker and the person dying is different as is shown by the varied usages. The self identifies itself in semantic space and is identified by others through relations expressed by language.

With this seeming paradox in mind—that the self is uniquely personal and thus constitutes the individual, but that it cannot be seen apart from symbolic interactions with the group—let me go on to other views of self. Jung would use the word "persona." From his viewpoint, persona is not the legitimate individual. "Fundamentally, the persona is nothing real: it is a compromise between individual and society as to what a man should appear to be. He takes a name, earns a title, represents an office, he is this or that. In a certain sense, all this is real, yet in relation to the essential individuality of the person concerned, it is only a secondary reality, a product of compromise, in which others often have a greater share than he. The persona is a semblance, a two-dimensional reality to give it a nickname."[8] Jung does go on to point out that in calling the persona unreal, he is being somewhat unfair because there is something individual in the peculiar choice and delineation of the persona. It is also clear from his description that more than one persona is possible, though he does not say that. For him, growth is the emergence of the individual as an adjudication between persona and the personal unconscious (as he sees it) in the process he calls individuation.

Perhaps for the Greeks there was no individual, totally personal (and essentially private in the Jungian sense), and for them the word persona did not represent a semblance but was indeed the "real" self. The modern sociological view of role most closely approximates the thing of which the Greeks spoke, as in the role of physician, child, mother, or even philosopher. As often used, and possibly to exaggerate slightly, I shall say that the Goffmanesque role seems to be for Erving Goffman pretty much the whole person. As used by me, a "self" is decidedly not the whole person, just as a "self" as I have portrayed it is not an unreal person. Nonetheless, in these two almost opposite views, Jungian and Goffmanian, both supported by experience, one can see why the word remains so fuzzy.

I suppose it is necessary to mention one other way in which the existence of more than one self has been handled in normal individuals, that of transactional analysis. There the existence of a child self, parent self, and an adult self are acknowledged and the aim of treatment is to bring to dominance the presumably healthy adult self. The immense popularity of transactional analysis derives, I suppose, both from the simplicity of its concepts and from their approximation to something recognizable within each of us. Although such concepts may be popular and lend themselves to group activities, amusements, or even effective therapy, they do not answer the essential questions raised by this essay: How are borders between selves maintained and how does adjudication between selves take place? And, ultimately, they cannot explain the conflicting ethical decisions raised by this essay in the first place.

For the purposes of this discussion, I have bracketed the domain of the unconscious. I am aware of its existence and I believe it is one of the determinants of the form and content of the "self" which I am describing. Indeed, in some people and under certain circumstances, it might be a major determinant of the expressions of self. But the relationship of the unconscious to the self is, to say the least, unclear; thus, I cannot deal further with it at this time and remain coherent about the self beyond saying that to postulate the self as I have done, without recognition of the fact of unconscious determinants, would be as barren as suggesting that the self is an entity solely determined by the unconscious. In one sense, the paradox of consciousness exposed by Unger applies also to the unconscious. It is clearly a uniquely private part of the individual and yet it has no coherent meaning in development and expression apart from others, and in that sense must penetrate consciousness. But the self is consciousness, and consciousness is more than an expression of the unconscious. In this discussion, for definitional purposes, the unconscious can also be bracketed and included in the realm of unknown and possibly irreconcilable conflicts, whereas the conflicts in ethical decisions of which I am speaking are reconcilable and result in action.

Acknowledging the unconscious and the fact of unconscious determinants in its constitution will make it impossible totally to subsume the self of which I speak under role. The self as role alone is also denied by the fact that the several "selves" in each person are authentic to that person, although someone may say of his actions, "I was surprised at what I did," or "I did not know I felt that eay."

I have been saying that in certain decision-making situations, different "selves" within the same person come into conflict. For example, the child versus the physician, where the child of a dying patient is also a physician. Or, as in the case of the woman flying to the bedside of her dying mother, the "adult intellectual" versus the "child." In that instance, the "adult-intellectual self," on her way to the bedside, seemed to contain or have access to information useful in deciding whether aging, sick parents should be "allowed to die." And this information translated itself into a formulated behavior or set of actions. In this instance, and in many others similar to it, the behavior did not arise out of previous experience with the situation, which would speak against the "adult-intellectual self" being a stereotypical behavioristic entity—for that person had never been in that situation before. Furthermore, information concerning, and "ideal" behavior toward, dying parents has changed in recent times (what is considered "right" in 1977 is different from what was considered "right" in 1967 or 1957). Thus, behavior is capable of change not subject to direct experience. With regard to information/behavior, two other things seem to be true; information/behavior does not seem to be especially tentative—people can act sure of themselves and also can be very sure of themselves about a reality they have never seen.

Now, however, the "adult-intellectual" self arrives at the bedside of her dying mother and inner conflict arises. That is, the previously determined decision about allowing the mother to die is called into question. There are several reasons for this. First, the reality of the situation is not as clearcut as was previously believed. Is the mother actually dying? It is suddenly apparent to the woman how little she knows about dying patients. Would a transfusion save the mother's life or merely prolong the process of dying? Would she be a bad child if she refused this to her mother? Was the doctor being a good human being in his recommendations, or "merely a doctor"? But, most importantly, the patient in the bed is suddenly not just a member of the intellectual category "dying patient," but "my mother who is dying." With that perception, the child self comes to the fore. For one thing, children do not make decisions, parents do (although role reversal between children and elderly parents commonly occurs). And, of course, a child does not want to lose a parent, and the suffering of a parent, more than that of anyone, is (usually) intolerable. Conflicts arise. The child wants it to be over for both altruistic and selfish reasons, but for it to end, the parent must die. Certain overriding rules come into play, such as "Thou Shalt Not Kill" and "Honor Thy Father and Mother." Lesser rules not strictly related to the situation also have force—rules of etiquette. It is not polite to continually question a doctor or make a scene.

Thus, the "adult intellectual" en route to the bedside of her dying mother finds, on arrival, that the reality of the situation is not nearly as clearcut as her previous well-defined beliefs about it, and that the person she has become at the bedside is no longer "adult intellectual" but rather "child." With that case in mind, let me reexplore the two similar cases in which the young physicians acted in an opposite manner, one in 1971 and the other in 1977. What had changed? I do not believe the role (in the strict sociological sense) of physician changed significantly, although there have been some changes of style. It is difficult to imagine that a change in the unconscious domain of the 1977 physicians was responsible. Patient-doctor and doctor-to-doctor etiquette is essentially similar now to that of 1971. The reality of the two cases was also the same: in one the young physician protested allowing the patient to die, although family and (apparently) the patient were for it, and in the other instance, the young physicians allowed the patient to die when family and patient wished life to continue.

At this point it is worth trying a tentative conclusion. It appears reasonable that the self is a rule-constituted entity and that between 1971 and 1977 the rules changed. Before going on to clarify what I mean by rule-constituted, I must discuss something that would merely trivialize the conclusion. Why say that the cases suggest that the self is possibly a rule-constituted entity instead of saying merely that what is considered good medical practice has changed? This is similar, for example, to saying that in 1971 it was good medical practice to give oral antidiabetic drugs for diabetes even if the patient wanted insulin, and in 1977 it is not (usually) good medical practice to use oral agents even if the patient wants them. Changes in medical practice in regard to diabetes are technical changes that result from new scientific evidence. The doctor's decisions are based on analytic thought—careful analysis of the situation and weighing of the relevant facts. The patient's wishes can be examined and rejected (or heeded) with an explanation offered to the patient in which all the factors can be made explicit. The issues of value are implicit and would probably not enter into the discussion because both patient and doctor probably agree (although not necessarily so) that it is more important not to be made sick by diabetes than to be merely happy about the kind of medicine one takes.

The cases of change in regard to dying patients are distinctly different. Here the explicit medical evidence did not play the largest part in the decisions. Indeed, on the medical evidence alone, the young physicians were wrong in both instances. Rather, in these cases, the issues were fundamentally issues of value. This kind of thinking is primarily not analytic but rather valuational, a kind of thinking, I have argued elsewhere, that is necessary to medicine but neither explicitly taught nor honored.[9] The evidence on which such decisions should largely be based is moral, not technical, and the source of such moral evidence (the wishes of family and patient) was avoided in both these instances. Thus, the conclusion that it is a change in rules that has occurred, not merely a change in medical practice, seems warranted. In defending the conclusion, however, it appears that the rules that help constitute the self are value-laden. Let me now discuss further what I mean by rules.

I admit at the outset that I am going to have difficulty explaining what I mean. Providing alternative words like precepts, principles, injunctions, or what have you, will not clarify the matter. The principal reason for the difficulty is the very wide range of rules that seem to constitute the self. First there are almost universal rules such as "Thou Shalt Not Kill," or "Honor Thy Father and Mother." Then there are rules that are implicit in professional roles, such as the requirement of physicians that they respect their patients' confidences. Some rules are culturally derived, such as the father sits at the head of the table, or the older brother will speak to the doctor. Rules can also be more personal and part of a family's habits, as in the way a woman arranges her refrigerator. Or they may be entirely personal, as in the way an individual may always precede an aggressive statement with a self-deprecating phrase or gesture. It is readily apparent that as one goes from the more universal to the most individual, what are being called rules appear less like rules than behavioral traits. What function is served by calling all of them rules? In the sense that they all determine the self as "separately identifiable in action," they are all precepts that guide the action of the self. The self is a particular and idiosyncratic mix of rules or precepts that are cohesive over time. The fact that some rules are universal and others almost totally unique to an individual confuses things. The hugely varied origins of the rules make consideration of self difficult and allows such argument between those who see individuals as role-sets and those who have a more existential view. Indeed, to use the word rule for "Thou Shalt Not Kill" as well as for my child's aggrieved cry, "No olives? We *always* have two kinds of olives at Thanksgiving," seems to trivialize the universal. In respect to their vastly greater importance, we call the universal rules moral laws and by doing so acknowledge the harm that comes from their transgression. No such harm arises when there are no olives on the Thanksgiving table.

I can understand that it may be disturbing to include within the same discussion universal moral laws and what seem to be merely habits. Indeed, if the purpose of this essay were to explore the origin of overriding goods in human interaction, the disquiet would be justified. But my purpose here is to explore the self and see of what it is constituted, and, further, to show that one can have more than one self and that these selves can come into conflict in ethical decisions. The rules or precepts of the self do indeed range over so wide a spectrum. But at whatever level, laws, rules, precepts, or habits are based on a belief or beliefs about the world; the very large world of all persons or the very small world of this individual physician or that "adult intellectual." Dorothy Emmet has provided considerable thought and discussion about the difficult problem of the relationships between the various classes of rules and precepts.[10]

Such a constitution of self is parsimonious or at least efficient because, it seems to me, the fundamental problem to be solved for the self is the constantly changing and uncertain nature of the reality in which decisions must be made. Where the self, faced with a decision, is able to determine merely what rules apply in (relatively) general terms, uncertainty is considerably reduced.

If we return to the "adult intellectual" en route to the dying mother, we understand why she is able to make a decision in an area in which she has no direct experience. She does have beliefs about the care of the dying, and those beliefs have been formed by the experience of others, recent readings, discussions, and so forth. Those beliefs have changed in recent years so that the former belief that one should keep the dying alive because "you never know when they will develop a new cure," changes into a belief that prolonging dying will increase suffering. The beliefs of young physicians have also changed between 1971 and 1977 and in the same direction. Where "keep alive at all costs" was the humane thing to do in 1971 (for young physicians), "allow to die" becomes the humane thing in 1977. If, as I believe, uncertainty is the problem to be solved for any self, we can understand why the young physicians acted, both in 1971 and 1977, seemingly without regard to the wishes of the respective parents and their families. As long as the situation conformed to their belief, a precept existed for action. Talking with the families and getting more information would not decrease uncertainty, but rather increase it. It would, for example, place the young physician at emotional risk, for (as it is said) identifying with the children (near his age) and thus seeing the case from the perspective of both a child self and a physician self could only result in ethical conflict. In just that manner, when the writer arrived at the bedside of her dying mother, both uncertainty and change in perspective occurred; she could not be certain whether the realities of the situation conformed to her "adult-intellectual" belief (i.e., was the mother really dying), and in addition, she became a child at the bedside of her mother.

Earlier, I indicated that between 1971 and 1977 the rules changed for the young physicians. Now I suggest that beliefs changed. Indeed, the rules may have stayed the same ("humane") and only the beliefs changed, or both may have occurred. However it may be, it is clear that rules and beliefs are intimately related. Beliefs, as Gunnar Myrdal has said, are ideas about how the world works. But if that is the case, then both rules and beliefs are intimately related to another vital element—perception. The problem of percepts and beliefs is similar to the problem of beliefs and actions and still revolves around the issue of uncertainty. When the person merely "sees" what existing percept (perceptual gestalt) applies to that which is perceived, less uncertainty is created and a decision can be made more efficiently than if a person had to constitute a whole new percept for each occasion. For a belief to remain intact, the perceived world must support that belief, and the belief must remain stable if action is to be unambiguous.

Self as perceiver and self as decider are intimately linked. It may be the case that different selves simply perceive the same situation differently. For example, the young physician perceives a dying old woman, the child perceives a dying mother. We all know of the classic blind men and the elephant, but blindness need not be literal blindness. "Rashomon" is a story whose interest lies in the totally different perception of the same events by different participants in the action. Once again, the possibility is raised that the rules for the different selves are all the same but the perception of events is different. As a physician at the bedside of a patient in pain, I am relatively insensitive to the pain and I do not suffer with the sufferer. As a child at the bedside of my parent in pain, even though I am a physician, I suffer.

Three possibilities now exist to account for what appear to be the existence of different selves within the same person. The first is that the self is a rule-constituted entity and the different selves are separately identifiable in action because each has different rules or precepts for actions (although clearly some or even many of the rules may be held in common). As Paul Ramsey has suggested, one would then have to see these selves as parallel strings of conscience which must ultimately come into conflict. An alternate possibility is that selves are constituted from different beliefs and that the rules may or may not be different. However, if the beliefs are at variance, different actions may result. Or, whether rules are the same or different and beliefs are the same or different, the selves are held as distinct entities by differences in perception. I am unable to distinguish which of the three alternatives is most likely. Whichever one might choose, difficulties arise. At the present I am unable to resolve the issue and I leave it reluctantly because I know how important it is.

Some have objected to the notion of self as I have portrayed it, believing instead that a person can be of "two minds" about something, or hold conflicting opinions about what to do, or simply be in conflict. I may be of two minds about which candidate to vote for simply because neither meets my single desire. But in the case of conflicting opinions, since most, if not all of us, are in conflict many times in life, I must ask whether the differing opinions within us are random occurrences? Clearly they are not. Opinions are rooted in beliefs. Each belief on which an opinion is based and even each rule and percept has a history. Each follows from a previous belief or rule or percept and the fact of conflict between opinions suggests a differing, nonrandom history. Then what holds these histories if not different selves within us? What more logically cohesive or parsimonious solution to the simple fact that we each hold often deeply conflicting opinions can be offered than the proposal that a person is constituted of more than one self "separately identifiable in action across time"? I can conceive of none.

No matter which, or which mix of, rules, beliefs, or percepts are the primary constituents of the selves, ethical conflict between selves is inevitable.

My original interest in this problem arose not so much from an attempt to show that there are different selves, but from a consideration of the fact that the same person can come into a situation in which he or she is in conflict about an ethical decision. I was interested first in understanding how the borders of the conflicting interests were maintained, and second, how adjudication between the conflicting selves takes place. For if action is to follow, adjudication must occur.

I believe it to be true, as Jung says and as Mead implies, that the process of growth, maturity, and ultimately of becoming wiser is a result of the multitude of resolutions between the different selves that constitute the person. Further, the process of becoming mature, if it occurs, involves learning the fact that adjudication can take place and how it is done, and how to tolerate the uncertainty that is inevitably involved.

Let us look at the original problem in a somewhat different way. The identifying characteristic of the actions of the young physicians in both 1971 and 1977 could be considered that of dogma or ideology—the application of rules and/or beliefs with a relative insensitivity to the actual facts of the situation. These dogmas or ideologies are systems of rules for action and perceptual gestalt (or systems of beliefs—for by this time, it is becoming clearer that no simple distinction can be made between rules, percepts and beliefs) often learned far in advance of the reality to which they are meant to apply. Two things are true of these ideologies. First, they are made up of relatively simplistic views of the world. That is, the beliefs seldom contain subtle distinctions or allow of many exceptions or extenuations. Second, it is characteristic of young adults that their different selves may be alarmingly sure of themselves (often inversely proportional to their knowledge of the world). Thus, it is in young adults that the concept of different selves "separately identifiable in actions but inseparable from the person" is most easily substantiated. Self as child with sure opinions about the parents, and at the same time self as young wife/husband or mother/father almost stereotypical in behavior (sometimes strikingly similar to the parent that the self as child criticizes) and self as young physician. Such ideological cohesiveness is quite functional in that it is a solution to the problem of uncertainty. But in the situations that I posited, where selves (and thus ideologies) come into conflict, even greater uncertainty is created and must be resolved.

The uncertainty, the contradiction is intolerable. However, although at this time I cannot give as firm an account of the resolution as I have of the problem and its genesis, I must try. Where can the resolution come from if not from allowing more information to enter the decision-making process. That information must derive from reperception of events or reinterpretation of the original perception. As more facts are allowed to enter, the problem that created the conflict becomes increasingly specific. In increasingly specific problems, the general rules and/or ideological beliefs that constitute each individual self become less useful as absolutes, serving only as guidelines. Indeed, some rules, some whole ideologies, and the self dependent upon them might conceivably crumble in the face of new information. That must be, when it occurs, a profound event. We hear people say, "After that happened, I was never the same again. I never saw the world the same way as before." There are images for this in all cultures—the phoenix, rebirth, true conversion, and others. For most conflicts no such deep change occurs, rather, the rules that cover the new circumstance are thought through, considered and reconsidered, until action results. The action serves to produce more information (since most acts occur over time) that reinforces or helps solidify the new interpretation.

But where is the original perception that is available for new interpretation or wider recognition? There must be something within us—an experiencer—able to perceive and store perceptions more "true-to-reality" and not as guided by unconscious determinants or conscious beliefs.[11] An experiencer whose fresh interpretations or understandings of the original experience can serve to adjudicate between selves. If this is so, then selves are not experiencers as much as assigners of understanding to experience. This may be the case because we often say of someone, "How can he believe that (or say that, or act that way) in the face of so much evidence to the contrary?" There the self plucks from experience a percept, a construction of the experience that best serves to maintain itself.

Ideally, however, with growth and maturity, the experiential increasingly dominates, uncertainty becomes more tolerable, and the distinctions between the selves blur. When and if this happens (for it is by no means necessary, but rather to be desired) the person that emerges and comes to dominance is an amalgam of the selves (though one self may persist to a greater degree than another). In this process, previously learned beliefs and their rules, even profoundly important ones, such as belief in God, come to be truly the property of the person. The paradox of consciousness—between the person as authentic individual and the person as only socially validated—is reduced as the individual becomes more essentially self-referring and self-made, and at the same time more at home (though not necessarily happy) with society and its symbols, less afraid of uncertainty and conflict.

What have I tried to show? Starting from situations in which individuals were in conflict with themselves over a. ethical decision, I have postulated that different selves within the individual are in conflict. At first it appeared that selves were rule-constituted entities, but quickly it became apparent that they are also constituted of beliefs. The distinction between rules and beliefs began to blur as though one could not clearly distinguish between them or say that they are truly separate. And then the place of perception in the maintenance of self became apparent as precept, belief, and percept merged as the constituents of self, serving the fundamental need of reducing uncertainty. Finally, resolution can be achieved when fresh experiential knowledge is allowed to enter, and in this process, occurring through multitudes of resolved conflicts, distinctions between selves blur and an individual emerges, authentic to itself and authentic in relation to its experiential world.

NOTES

1. Eric Cassell, "Preliminary Explorations of Thinking in Medicine," *Ethics in Science & Medicine,* vol. 2, pp. 1-12.
2. Eric Cassell, "Permission to Die," *BioScience,* August 1973.
3. Stephen Toulmin, "Self-Knowledge and Knowledge of the Self," Ms.

4. George Herbert Mead, *Mind, Self and Society*. (Chicago: University of Chicago Press, Phoenix Edition 1934), pp. 166, 167, 171.
5. Ibid., p. 173.
6. Toulmin, "Self-Knowledge and Knowledge . . .," p. 30.
7. Roberto M. Unger, *Knowledge and Politics*. (New York: The Free Press, 1975), p. 215.
8. C. G. Jung, *Two Essays on Analytical Psychology*. (New York: Meridian Books, 1953), p. 167.
9. Cassell, "Preliminary Explorations of Thinking in Medicine."
10. Dorothy Emmet, *Rules, Roles and Relations*. (Boston: Beacon Press, 1966), chap. 4, 7, and 8.
11. Eric Cassell, "The Subjective in Medicine," *Clinical Judgment* (in press).

Commentary

Response to Eric Cassell

Paul Ramsey

ERIC CASSELL gives us an account of "selves separately identifiable in action and across time but inseparable from the whole person." He understands self-conflicts in medical and moral decisions and actions to run so deep that "being of two minds," "persona" or "roles" are inadequate terms by which to express the divisions experienced within one and the same individual. Instead, he suggests the existence of different "selves" separately identifiable as the agents of their deeds, each authentic to the person they constitute across time.

In the end, I shall want to say more about the person within which these several selves seem to reside. "The whole person" remains strangely inert or silent in Cassell's drama of adjudication, resolution, and emergent maturity. Here, if anywhere, the responsible self is to be found, even if all that can be said is that a person in such conflict has—by Cassell's reckoning—a heavy burden of self-containment and an enormous moral task in ever coming to self-realization.

Most suggestive and instructive is Cassell's descriptive analysis of the entity selves who are for him the chief actors in decision-making. We should not quibble, in my judgment, about the uncertainty he expresses about whether he has the precise terms for a self's constituent elements (beliefs, rules, and percepts) or over which of these ingredients to accent. This is progressively clarified in his chapter. The troika may be led by either beliefs, rules, or percepts; these are alternative possibilities. No simple distinction can be made between these elements. I judge that they are in constant interplay, with settled priority ascribed to none. The portrait presented is that of selves, each of which is a particular and idiosyncratic mix of rules, beliefs, and percepts that are cohesive over time. It becomes quite clear that there is a symmetrical and everchanging interrelation among these vital moments in a self's internal formation. Thus, Cassell conveys a sense of the wholeness of each of the selves he depicts, even if he fails to glimpse the wholeness of the person who is responsible in, through, and for them all. The whole person is a quite silent partner, or else it is postulated as an emergent.

Each aspect of one of the separate self's history and internal formation is worth brief examination. A self is a rule-constituted entity. It is also a belief-constituted entity and an entity constituted by its percepts of a situation. It is evident that the reality to which these terms point is a unity of diversity for a self's lifetime. An aspect of a self to which one term points interpenetrates the other aspects also, so much so that Cassell's language is frequently interchangeable. I count this to be a strength of his analysis. The "rules" changed between 1971 and 1977, as did the rule-constituted physician selves. So did the "ideal" behavior toward dying patients, or what was considered "right." What is the latter if it is not a change in the belief systems and of the belief-constituted selves? So "rules and beliefs are intimately related," as Cassell says.

Perception is also intimately related with them. The "adult intellectual's" contemplated behavior came from no previous experience at the bedside of the dying. Her perception of such situations came from reading and similar sources, even as the 1977 young physicians may have undergone change not subject to direct experience. Here we have selves fairly free of Mead's origins.[1] Cassell's selves have belief and rule systems penetrating percepts. Conflict arises when the child self realizes how little *she knows* about dying patients. Thus, percepts of the situation can affect knowledge and may alter belief or rule systems, or vice versa. So much so that any opposition between "mere information" and "a different self" disappears. For the several selves' different perceptions of the same situation garners different information about it. Their perspective upon the same information means that they are informed by different perceived worlds. Indeed, the 1977 rule-constituted physician selves perceived the patient (or did not perceive the patient) by the rules. I would therefore express the intimate relation between perception and the rule-constituted self or the belief-constituted self by reversing Cassell's statement that "for the belief to remain intact the perceived world must support the belief and the belief must remain stable if action is to be unambiguous." It is *also* the case that the perceived world remains intact so long as the belief remains and the self constituted by its remains stable.

In passing, two words of praise for Cassell's account of rules are in order. (1) His depiction has amplitude, all the way from "overriding rules" and "universal moral laws" to behavioral traits or habits. In some contexts we may need a narrower usage of the word "rules." But Cassell's account has the virtue of not splitting an ethics of *action* apart from an ethics of moral *character*. (2) His view has amplitude also in that possibly universal moral rules are included. These seem to me to be correlated with "a belief or beliefs about the world" and also with perceiving the world "beliefully." Again I suggest that selves are now so distant from Mead's theory of origins that we can question the extent of its explanatory value. Cassell's limitation of the range at one point to "the very large world of all men or the very small world of this individual physician or that adult intellectual" is either a slip or a mere prejudice—since elsewhere he allows that "beliefs and their rules" may include "profoundly important ones such as belief in God." Such beliefs may "come to be truly the property of the person." This is to say, a self may be constituted by its belief in God, with a corresponding rule system (character no less than "divine commands"), and so perceive the world in this light. That is an even larger world with which to perceive all small worlds.

Here a small objection arises, at least for this reader. Perhaps we need to challenge the implicit assumption of the normalcy of personal moral progress which I find in Cassell's essay. By that I do not mean that he would deny the reality of moral deterioration or breakdown. It is natural enough, moreover, for a physician ("older physician") to assume that more wisdom will die with him today than when he was a "younger physician," or tomorrow rather than today. Still I ask: Is not this the myopia of every present self? In the concatenation of selves, is there not always loss as well as gain? And in the absence of a general theory of moral agency, no way to tell gain from loss to be drawn from the ebb and flow of selves across time? I imagine I had worths as a child self (and not only because of worth my mother's love discerned) and as a student self that must be counted as losses now that I am a professor self; and praiseworthy endurance as a father self I would not care to have in actuality or to exercise again as a grandfather self.

Besides, how do we prove from experience that a self that manages to encompass a former self by subsumption into itself is better than a narrower, more concentrated moral resolve? Why should a norm of maturity be disguised and incorporated in the finding that "the paradox of consciousness—between the person as authentic individual and the person as only socially validated—is reduced as the individual becomes more essentially self-referring and self-made, and at the same time more at home (though not necessarily happy) with society and its symbols, less afraid of uncertainty and conflict"? St. Francis was an individual who was scarcely at home in the late medieval world with its growing towns and money economy, or with his merchant father. Thereby he was doubtless more essentially self-referring and self-made; and, fearless of uncertainty and conflict, constituted himself with a belief system and a rule system that was a new creative constellation which influence from the Christian past cannot fully explain. St. Francis was authentic in self-possession and self-reference. Doubtless he also was in tune with the world he perceived from the outlook of his belief system. Still by some measures his was a narrow resolve. Certain nuances in Cassell's chapter suggest, to the contrary, that maturity in the pilgrimage of an individual soul in the making must mean increasing integration of increasing diversity—which was Spencer's "Darwinian" definition of social progress.

My crucial, critical response to Eric Cassell, however, is as follows. Assuming the conflict of the selves he depicts, he asks: Where can resolution of these deep-seated contradictions come from? The selves in conflict are only "assigners of understanding of experience." What can adjudicate between or among them? Cassell posits as the source or trigger of resolution—leading to an emergent, integrated self—something he calls an "experiencer." This cannot be one of the selves in conflict, nor is it "the whole person" who remains, so far as I can see, a nonfunctional battlefield over which several selves are at war. Indeed, Cassell describes this "experiencer" rather like a computer: "able to perceive and store perceptions." "Where can the resolution come from," he asks, "if not from allowing more information to enter the decision-making process?" Clearly the "experiencer" is a sort of transmitter-receiver station needed to account for the emergence of a more integrated moral agent. If the experiencer is not a separate organ, located among the conflicted selves and also within the whole person who is their placid common residence, it does the work of such an organ. It is a capacity for reperception and storing new information.

This resolution of selves in conflict in medical and moral decision and action comes as a major disappointment and as narrowness of vision when tested against the richness and profundity of Cassell's own account so far. He has, of course, made room for the possibility that new information—new percepts—may cause belief systems and the self dependent upon them to crumble, and for the possibility that rule systems can become only lamed guidelines or traces of a former self. But, I ask, has he not also allowed that a belief system—including belief in God and in universal moral rules—may have controlling effect upon how the world is perceived? The constituent elements of each self were in dynamic interrelation: beliefs, percepts, rules, and traits of character. So Cassell denies that "new information" explained the difference between the 1971 and the 1977 physician selves. He also rejects the bifurcation between cognition and affect.

Here, then, Cassell falls into a major contradiction. He has offered no reason why a self's belief system revivified or its adherence to universal moral principles may not take the lead and alter the relevance and importance of information received through perception—indeed, alter perception itself. Nor for that matter can I see any reason to exclude, for good or ill, the possibility that a rule-governed self (which includes trivial traits like orderliness) may not decisively influence the information received and how an individual learns to tolerate uncertainty. By Cassell's own account, any of these sources could be the motive power behind a resolution and adjudication of selves in conflict. I can only conclude that the author nodded—and lapsed from his own more profound analysis—when he settled for empirical perception and new information as the sole source of a person's movement toward integration. If I am correct in saying this, he yielded to the going assumptions of the present age and its presumed "scientific" mentality. Before that lapse he had given us better foundations for moral decision-making. Cassell should dismiss his reified "experiencer" from its usurped sovereignty.

I now wish to lay down a supplement alongside Cassell's descriptive-analytic account—a *needed* supplement with which Cassell, I have reason to believe, would not altogether disagree even though he has not chosen to revise his chapter or otherwise to incorporate my proposed enlargement of his expressed viewpoint. My supplement proposes an active role for "the whole person" or "the same individual" within whom Cassell locates his conflicting selves. This is the self of all those *Menschen in Widerspruch* that Cassell discovers, whose moral task and responsibility is their adjudication and resolution. That person, whom Cassell mentions repeatedly but who remains inert and silent in his drama, is the experiencer, the believer, the rule- or trait-governed initiator of its emergent integration, if such integration takes place.

I mean to subscribe to no particular metaphysical, ontological, or psychological theory about the nature of the whole person, the subject, or the entity—if such it is—that seeks and hopes for Cassell's emergent wholeness, whose task is the resolution of conflict between selves. Kant would have called the whole person an "ideal of pure reason" beyond the limits of cognition based upon perception alone, or else the "transcendental unity of apperception" that must be presupposed for there to be experience of a world. These unities are allurements ahead or behind all endeavors to find in experience the whole self; they tempt us to suppose we can accumulate cognition of the subject who is morally accountable. At this point, as is well known, Kant posited a "noumenal self." I do not here do so.[2]

But I do suggest that we can validly use terms ascribing moral responsibility (praise and blame, regret or guilt) across time which are more extended than most of Cassell's selves. The validity of these moral notions points to what I shall call only a liminal sense of self (to avoid any metaphysical, ontological, or theological theories about it) that is the unknown and perhaps unknowable subject behind or beneath the conflicting selves Cassell so vividly describes. Such a "whole person" is presupposed by the moral language we rightly use when we say no more than that the selves-in-conflict are each authentic to the person they tear asunder. For there to be moral conflict *some sort* of inostensible identity (and unease) must necessarily be presupposed across the lifetime of Cassell's selves connecting them. "To tolerate uncertainty," or to find it intolerable, admits an inward action on the part of the individual person caught in conflict of selves.

Our moral language certainly does so.

Praiseworthiness or blameworthiness are not as fragmented as Cassell's selves seem to be. Murder, or being a willing and principal self in the Holocaust, are not the only wrongs for which there is no statute of limitations, however understandable is the self or behavior that actually came to replace the self that did those deadly deeds. We do not cease to be the one (the whole self in some sense) who deserves blame for a wrong, injury, or neglect done long ago by becoming a self that now would never, never do such a thing, and can scarcely understand the self that once did that harm. Nor is a hero any less *that* hero because he becomes a blubbering slob after the war; or a year of saintliness less attributable because—as it turns out—martyrdom would have been easier to live up to in later life.

Consider the case of the young physician house officers who failed to provide a respirator for Esther R. Begin by asking, what are we, as moral agents, doing making intellectual grist out of that failure? As I understand the case, these young physicians not only refused to carry out the orders of the chief attending physician, they also acted directly against the unambiguous request of both the patient and her family, and denied Esther R. something she had every right to ask: to be helped to go home with a belt respirator. They "stood idly by the blood of their neighbor." That verse from Leviticus is the basis in Jewish law and rabbinical teachings (honored certainly by both orthodox and conservative Judaism) which holds every moment of life to be the same as an eternity, and so finds no basis for distinguishing (as in Christian ethics—which is softer on "respect for life") between omission and commission or between allowing to die and killing. I rather hope "Esther R." had a good Jewish surname, like Rabinowitz. In that case, these house officers imposed on her a shortening of life that she (if religious and instructed in the moral teachings of her faith-community across time) would not have been able to consider even as an option. Cassell is a physician well known for his view that medical judgment itself is valuational, and for his view that this fact is too little honored. Here we must add that to treat a patient as a person means not only physician valuation but also his respect for the patient's valuations—including Esther R.'s possible religious beliefs.

What was done to Esther R. can scarcely be described as treating her as a person. In the first paragraph of his chapter, the author states that such a standard is insufficient without providing a wider understanding of the concept of a person. That, of course, is correct—although how to get at the concept and how much instruction a physician needs about it in order to know its claims upon his practice may be debated. At once, the theme shifts from treating the patient as a person to a discussion of physicians acting as persons (or selves) in treating patients— presumably as persons. *They* are the ones who "valuate," and this sets the stage for the cases of Dora S. and Esther R., with their graphic demonstration of a shift in physician and societal values between 1971 and 1977.

Now, undeniably all this is worth giving attention to. To do so is without doubt among the tasks of intellectuals, and important insights about sociocausal factors may flow from it. Still I for one fail to see what the ensuing descriptive-analytic enterprise has to do with the foundations of ethics or, in particular, how it can enlarge by one square millimeter our concept of moral agency or of the responsible self.

For that, we would have to launch a discussion of the reasons for saying that the shift of young physician valuations from 1971 to 1977 was a moral decline. Or of the grounds for at least saying that, while the shedding of routinized physician care in the 1971 case of Dora S. was some sort of a gain, its replacement by physician judgments in the 1977 case of Esther R. was a clear loss in physician persons treating patients as persons. Cassell says as much when he writes that in both instances the individual nature of the patient did not seem to enter the decision. In the case of Esther R. it entered not at all, or only as yet another sort of routinized imposition (like not treating a ninety-year-old man *because* he is ninety). In both cases, Cassell writes, "the evidence on which those decisions *should* largely be based, is moral, not technical, and the source of such moral evidence (the wishes of family and patient) was avoided in both instances" (italics added). That *verdict* seems to me to be in no way weakened by Cassell's *explanatory* account that "the rules had changed" for the young physicians between 1971 and 1977. Yet, if he has fully explained those selves, the explanation is sufficient to excuse. Where and who is the self who is a proper subject of moral blame for yielding to these temporal changes in the ethos; or, alternatively, the self to be praised for moving with the times?

We would have to launch a discussion that searched for the systematic changes in professional training or in our hospital system needed to slow down the acceleration of the depersonalization of patients and of doctors. We would have to debate whether those house officers ought not to have been fired or severely reprimanded or sued for malpractice or criminally charged or threatened by Aesculapius. We would have to ask what might evoke a change in them as moral agents (as physician selves), and works mete for repentance. Cassell explains that uncertainty is the problem to be solved by any self—to solve internal conflict or defend against it. The "ideological cohesiveness" of the 1977 selves was "quite functional in that it is a solution to the problem of uncertainty." So Cassell *understands* those house officers quite well—perhaps too well. Critical understanding may serve, of course, to help dissolve the defenses of ideological cohesiveness and clear the path for an emergent integrated moral agent. Still, the question to be pressed is: Of whom is it true *now* to say that such cohesiveness led to wrongdoing? Can one validly think of telling those ideologically cohesive selves to take their problem out of the setting where dying patients are treated? Is there room in Cassell's account for the person across time as a responsible self?

My point here is simply to suggest several, perhaps banal, questions or approaches that have at least the virtue of presupposing moral agency and responsibility on the part of a person. I do not see how Cassell's account of the decision-making of selves in conflict, twisting and turning slowly in the social winds, gives us a basis for ascribing moral responsibility to any particular self. Presumably, the 1977 house officers would say that the 1971 physicians did, or would have done, wrong. Suppose one is the same person—still young in heart. Presumably s/he would say that s/he did wrong. To what self does "s/he" refer?[3] Where is the self that endures through all change enough to look back on a succession of selves with even a mild sense of properly placed *regret*?

Perhaps we could stretch Cassell's "experiencer" to mean "the whole person," "the same person," the believer no less than an information-gatherer, the moral agent, the self behind all selves that does not appear in their overt conflicts. I would say that "I" often or usually make contradictory moral or value decisions depending on the circumstance. But then I am liminally aware that the contradiction is not authentic, not *mine*; yet also liminally acknowledge that these opposed decisions are authentic, that "I" make and made those decisions. If, when one of my selves was, I was not; and when I am, that self was not (like death to the Epicureans), where then is the contradiction of moral agents (selves)? I either am responsible for all those selves or I am not. If I am, then I am the self behind all selves Cassell has aptly described. If not, I am not morally accountable for them, and they certainly are not accountable for themselves or for me. Moral notions have no meaning unless I am "the same individual" in some sense that matters more than Aristotelian matter or than a biological organism in which conflicts between selves goes on.

I am asking what Cassell wants to say about what I call, for want of a better term, the liminal or penumbral self within which the activity he aptly describes goes on. One cannot have a battle, or a victory for maturity, without a battle*field*. Cassell uses stronger language when he speaks of "the whole person." Yet this person plays no role in his account of moral agency. Cassell needs a liminal self energized if he is to explain "what maintains the borders" and "how adjudication takes place" from which maturity might follow. Adjudication seems to me to be impossible without a latent adjudicator, or maintenance of the borders without a border guard. If we agree to refer simply to a self on the horizon of the several describable selves, who is no *where* in identifiable particular and at no particular time because across all times, this would be to move in a Kantian direction. I go no further in that direction than well-placed *regret* for a past self requires.

These seem to me to be unanswered questions of some importance. Perhaps I am making a mountain out of a molehill or—to say the same thing—a grave philosophical issue out of some very simple truths. Perhaps Cassell did not mean to raise these questions, and simply presupposes that *the* person is there all along in some sense overseeing the behaviors, the selves, and the outcomes.[4] Then that, I would suggest, is the responsible self of whom we can best speak in the same breath with talking about moral agency and its norms. Person$_1$ is a necessary presupposition if morally mature Person$_2$ is to emerge—rather like the primordial and consequent nature of God in Whitehead's philosophy, where also process reigns.

NOTES

1. George Herbert Mead's *genetic* account of the elicitation of self-awareness or personhood may be as good as any other, or better. By genetic account, I mean a true genesis; I do not mean to derogate causal explanations. Still, it is only a genetic or pedagogical account to say that a person's "experience of a self is one which he takes over from his actions upon others. He becomes a self insofar as he can take the attitude of another and acts towards himself as others act." (Those two sentences do seem to me to say quite different things.) This account of the transmission of the *achievement* of self is entirely credible. But such a history of human personhood is unintelligible unless normal "products of human generation" (traditional Catholic language) or "the same individual" (Cassell's language) have an inherent potentiality for self-consciousness, for "internalizing [its] conversation" with others. Otherwise, to describe "the true origin and foundation" of selfhood and of all "thinking" as "social" just as readily proves that the *Zeitgeist* changed its opinion (its self) between 1971 and 1977. The need for a central nervous system whose brain events are isomorphic with those supposed men-

tal events or selves is a pure prejudice. The *Zeitgeist* may have been coming to self-consciousness of another discrete "selfhood" through those house officers in 1977. There, if anywhere, lay the "moral agency" having some liminal identity across time and through the proximate "selves" of those young physicians. I do not see how Cassell can avoid these incredible suppositions if he does not make use of some concept of the self-identity of moral agents across time.

Indeed, Cassell's quotation from Roberto Unger seems to me to affirm that self-consciousness must be more than an inherent potentiality; it must be actualized in an individual if he is to recognize the gestures that come from another who is, like himself, a person. "These intentions can, in turn, be understood, because they are intentions you too might have," Unger concluded. "It follows that consciousness always presupposes the possibility of viewing other persons as selves that could, under favorable enough circumstances, see what one sees and believe what one believes." (Perhaps the genesis of selfhood takes place through the medium of language and the symbols of a culture; however, some philosophers argue that self-awareness and language do not have the tight connection Cassell supposes to be obvious.)

Thus, on the one hand, a self comes to self-consciousness by taking the attitude of another and acting toward himself as others act, by internalizing their conversation with him (Mead). On the other hand, a self could not know the other to be taking any attitude or acting toward him, intending any gesture, if it did not first have an interior understanding of the meaning of these personal moments, which are then ascribed to others (Unger).

2. At one point Cassell takes from me the alternative that his selves are "parallel strings of consciousness that must ultimately come into conflict." I said rather that one alternative to account for the self-identity of a moral agent across time would be to move in a Kantian-Tolstoyan direction and say that the ultimate subject of all experience and of moral discourse (not itself a piece of its own experience or a cognitive object) is *noumenal*—like a string that we never see along which stages of consciousness or selves are strung like beads with no interstitial spaces where "the whole person" can be viewed or objectively "experienced." The image of the beads and string was taken from Tolstoy's *On Life, The Complete Works of Count Tolstoy*, vol. 16, trans. Leo Weiner (Boston: Dana Estes and Co., 1904). The argument of *On Life* encompasses both Kant's *Critique of Pure Reason* and his *Critique of Practical Reason*. What Tolstoy calls our "animal personality" (which includes consciousness and all of Cassell's selves) is the same as Kant's "phenomenal self," while "ra-

tional consciousness" or our "special relation to the world"—the string along which the beads of ordinary consciousness or selves are strung—is the equivalent of Kant's "noumenal self."

3. The reader will note my clever way of preserving, in this age, the integrity of the English language when *read aloud*.

4. The first and still greatest psychologist of the Western world, St. Augustine, knew the problem of selves in conflict. This he expressed in his doctrine of the divided will (*Confessions* 8). Yet the person who willed and "nilled" the same thing at the same time, and could not yet will anything entirely, was continuous and active in that human drama. The self's wholeness came to light in *memory*. Memory, for Augustine, was more than a stream of cconsciousness sustaining continuity of selves across time. It was an *activity* of the soul, and e.g. (which means ex. gr.: *ex gratia*), a creative source of self-formation. Two passages from Augustine's *Confessions* (10: 8 and 40, Loeb edition) deserve quoting in a volume of contemporary essays, such as this, on the foundations of ethics:

> I demand that what I wish should be brought forth, and some things immediately appear; others require to be longer sought after, and are dragged, as it were, out of some hidden receptable; others, again, hurry forth in crowds, and while another thing is sought and inquired for, they leap into view, as if to say, "is it not we, perchance?" They I drive away with the hand of my heart from before the face of my remembrance, until what I wish be discovered makes its appearance out of its secret cell. Other things suggest themselves without effort, and in continuous order, just as they are called for—those in front giving place to those that follow, and in giving place are treasured up again to be forthcoming when I wish it. All of which takes place when I repeat a thing from memory.

> I, who went over them all, and labored to distinguish and to value everything according to its dignity, accepting some things upon the report of my senses, and questioning about others which I felt to be mixed up with myself, distinguishing and numbering the reporters themselves, and in the vast storehouse of my memory investigating some things, laying up others. Neither was I myself when I did this (that is, that ability of mine whereby I did it), nor was it Thou, for Thou art that neverfailing light which I took counsel of as to them all, whether they were what they were, and what was their worth; and I heard Thee teaching and commanding me.

Part III

The Scientific Roots of
Morals

9

Natural Selection and Societal Laws

Richard D. Alexander

I believe that modern opposition, both overt and cryptic, to natural selection, still derives from the same sources that led to the now discredited theories of the nineteenth century. The opposition arises, as Darwin himself observed, not from what reason dictates but from the limits of what the imagination can accept. It is difficult for many people to imagine that an individual's role in evolution is entirely contained in its contribution to vital statistics. It is difficult to imagine that an acceptable moral order could arise from vital statistics, and difficult to dispense with belief in a moral order in living nature. It is difficult to imagine that the blind play of the genes could produce man.

—George C. Williams

Introduction

Ever since Darwin published his *Origin of Species* literate people have tended to regard the attributes of living things as outcomes of an evolutionary process, and to suppose that humans like other organisms are in some way derived through organic evolution. In terms of searches for explanation there are several meanings to this remark. Laboratory scientists, for example, may assume that common or similarly derived mechanisms underlie physiological phenomena observed even in widely different animals, and they may as a result use simpler or more easily studied organisms, such as rats, to help understand complex species, such as humans, which are difficult to study in the laboratory. Primatologists may assume that humans are similar to other primates because of genetic similarity or a recent common ancestry; this assumption may be used in developing and testing theories about human behavior and its history, or about the phylogenetic derivation of humans. Paleontologists assume that the phylogenetic patterns they are able to trace across geological time are the result of mutation, selection, and isolation, as observable today, even though they are hardly ever able to reconstruct the environments in which the changes occurred well enough to understand the adaptive significance of ancient trends.

Biologists with a primary interest in the evolutionary process may seek generalizations about the nature of adaptation and extinction from the workings of that process now and in the immediate past, and develop and test hypotheses from this approach on whatever extant organisms seem best suited. They assume that if adaptation is a continuous process, and if generalizations can be extracted about adaptiveness of attributes common to many or most organisms—sex ratios, senescence, parental investment, sociality—then evidence of cumulative adaptive changes should appear in comparative analyses. At some point, with sufficient understanding of the evolutionary process and enough information about past environments, phylogenetic and other patterns from fossil evidence may be put together with adaptational information from the study of living forms in order to develop, as far as possible, both the actual reasons for long-term history and a greater predictability about the traits of living organisms.[1]

Here I wish to utilize the approach of generalizing and predicting from the process of adaptation as a vehicle for studying human sociality, especially to consider the nature and probable background of societal laws, norms, and traditions. The approach, as such, is not new. Since Darwin, it has been commonplace, at least in the Western world, to think even of human attributes in terms of natural selection. The reference, however, may be fleeting, only half-serious, or even principally humorous: "Ah, yes," we say with a smile, "Survival of the fittest!" Indeed, it is probably fair to say that in recent years most allusions to any relationship between human behavior and natural selection have been in jest. This may be partly because we are skeptical of the implication that what we do from day to day really has much to do with a process as simple and genetic as "survival of the fittest." It may be partly because we share the fears of others that evolutionary explanations imply an intolerable, robot-like determinism quite inconsistent with our views of our usual day-to-day, consciously planned, individualized existences. It may be partly because we sometimes recognize an irony in our identification of the "fittest" after the fact, because we know very well that we could have made it happen otherwise if we had intervened. It may partly stem from the suspicion that the purveyors of an evolutionary philosophy of life have some new brand of "social Darwinism" in mind—some notion that if natural selection lies behind human attributes then variations among people must be accepted as occurring along axes appropriately judged in terms of better versus worse, with such variations to be either allowed or caused to disappear by our deliberate contributions to the operation of some kind of "natural law" of survival of the fittest.

For these various reasons, applications of natural selection to human phenomena, and in fact as crucial explanations of biological phenomena in general, have drifted into neglect or at least a lack of centrality. This has happened despite unswerving recognition within biology that natural selection, or differential reproduction, is the principal guiding force in the evolutionary process.[2]

Some of the error and vagueness exists for unjustifiable reasons. Thus, genetic determinism is not in any sense a concomitant of the application of a selective model of human history. Genetic determinism implies long-term, irrevocable causation, but the elaboration of ontogenies and phenotypes which characterizes the evolution of life through natural selection is actually an opposite trend. Phenotypes and ontogenies—and especially behavior as an aspect of the phenotype—represent flexibilities that are opportunistically and strategically realized in the variable environments in which the organism lives out its life. In a sense, the most deterministic aspect of life is actually in the consistencies of the environments in which successive generations develop, and of course humans are masters at altering their environments far outside the limits represented by history. The existence and elaborateness of learning testifies to the absence of long-term causation in the environments of life, and therefore to the value of reliance upon immediate contingencies, especially in regard to social events. It is a curious fact that when genes are brought into the equation Genes + Environment → Behavior, as they must be, there is a widespread tendency to assume that the role of the environment is thereby necessarily underplayed and that of genes overemphasized. Even if this has been true in the past we have no choice but to leave genes in the formula and try to discover their role as well as that of their environment.

Similarly, there is no excuse for extrapolating from natural history to the development of value judgments about human attributes expressed in the present or future. Again, the opposite is more reasonable: To understand the past is not to bind ourselves to it but to deliver ourselves from its grip. Knowledge of the history of our own evolution should place us in the best possible position to cause the shaping of our future by human design, which in itself is inevitable, to proceed in desirable rather than undesirable directions.

I think, however, that fears of determinism and social Darwinism were not the real reasons why evolution, in the decades following Darwin, seemed to drift away from the front lines of biological investigation, and especially behavioral analysis. If evolutionary models had worked, they would have remained in the forefront. The reason they did not work was never squarely identified until George C. Williams published a dramatic refinement of Darwinism in his 1966 book *Adaptation and Natural Selection: A Critique of Some Modern Evolutionary Thought.* Williams's argument was that biologists had never clearly answered the question: Survival of the fittest *what*? For various reasons, they more often than not had assumed that the attributes of organisms have evolved because of their value in perpetuating the species, population, or social group. Williams showed that when the directions of change so indicated are contrary to those that would contribute to the survival of the genes of the individual organism, under nearly all conditions the latter will prevail.[3] What survives are the most reproductive sets of genes, and, concomitantly, the phenotypic potentials they have yielded in the environments of history (hence, that they may yield in the environments of the future).

This crucial refinement of Darwinism has generated a remarkable new surge of attention to evolutionary explanations, especially of behavior; and it promises to place natural selection once again in the forefront of every kind of biological investigation, including our efforts to understand ourselves through our history. Numerous authors have pointed out how this change in our thinking has rendered basic concepts and approaches obsolete in biology in general, and particularly in population genetics, ecology, ethology, and anthropology, because they depended upon selection operating principally at the population or group level.[4,5]

Darwinism, then, or the principle of differential reproduction, is a statement about why things are as they are with the entire world of life. It leads unmistakably to an attitude about the ultimate causes that lie behind the proximate mechanisms with which we, as individuals, must deal in our everyday, practical existences. Anyone who would challenge the philosophical implications of these statements must attack Darwinism at its base, which is to say, the entire idea and the universality of natural selection. For, as Darwin noted in 1859, to find an exception, an adaptation derived from some effect other than the cumulative influence of differential reproduction of variants, would not merely weaken his theory or reduce its overall significance but annihilate the entire idea. In the absence of any such challenge, we are not free to ignore the consequences of an evolutionary process, the cumulative effects of a continuing process of differential reproduction, upon all of natural history, including our own current attributes and tendencies. We are not free to deny such effects by assertion. We are not free to require that each scholar return to a defense of Darwinism before he develops a thesis on the assumption of its validity. In fact we are not free— any of us—to reject the evolutionary view of life. Logic, fact, and the absence of substantive challenges or reasonable alternatives clearly deny us any such frivolity. Whether we like it or not, and whether or not it has been a part of our personal educational, philosophical, social, or ideological backgrounds, we are required to accept that *background explanations for all activities of life, including our own behavior, will eventually be found in generalizations deriving from the cumulative effects of an inevitable and continuing process of differential reproduction of variants.*[6] The only question is their nature and the directness of their applicability.

From these introductory comments, I now proceed toward the particular discussion proposed for this chapter, the probable background of societal laws. Since laws are functions of societies, I shall do this by first discussing the probable backgrounds of societies and sociality.

Origins of Sociality

Once biologists had recognized that differential reproduction cannot easily be invoked to explain any attributes of organisms supposed to be good for the species or group as a whole, it quickly became commonplace to assume that attributes of organisms have evolved because in the past they helped the individual organism to maximize its reproduction. This interpretation has proved to be extraordinarily powerful in solving long-standing problems about altruism, population regulation, sex ratios, senescence, sexual dimorphism, parental investment, breeding systems, menopause, length of juvenile life, and a great many others.

The greatest impact of this revolution has been felt in the study of social behavior. Sociality can only exist in group-living organisms. Supposing that organisms do things because, in historical terms, they thereby help their own personal reproduction not only raised questions about all of the behavioral expressions commonly regarded as "social"—like cooperation, sharing, and all forms of altruism—but also changed our attitude toward voluntary group living. If groups were seen as forming, and the individuals within them interacting and cooperating, solely to help perpetuate the species, then deleterious consequences to individuals were to be expected and would not necessarily be minimized. Only the success of the group would be relevant. On the other hand, if behavior evolves to help individuals, we are suddenly aware that group living entails automatic expenses to individuals, such as increased competition for all resources, including mates, and increased likelihood of disease and parasite transmission. Accordingly we are made to wonder—for the first time, really—why animals should bother to live in groups. Why be social, beyond what is required to mate and raise a family? If the answer is that individuals living in groups reproduce *more* than individuals not living in groups, then, again for the first time, we are led to seek out the specific benefits that accrue from social life.

A few years ago, in the wake of this realization, I attempted an exhaustive list of such benefits for all organisms. To my surprise I was able to generate only three broad categories: (1) predator protection, either because of (a) group defense or (b) the opportunity to cause some other individuals to be more available to the predator; (2) nutritional gains when utilizing food, such as (a) large game, difficult to capture individually, or (b) clumped food difficult to locate; and (3) simple crowding on clumped resources.[7]

To my knowledge, no other reasons for group living, commensurate with the now well-established view that selection operates principally at and below the individual level, have been generated. It is a crucial point. Because individuals should tend to move apart, avoid competitors, and be nonsocial, large groups should appear only (a) when the resources of reproduction are so clumped that there is no alternative to close proximity (3, above) (a situation implying no cooperation, hence no special social organization), or (b) when cooperation contributes to individual reproduction in the population or species at large because of some extrinsic hostile force (1 and 2 above).[8]

Applying these hypothesized explanations to familiar organisms yields some interesting and surprising suggestions. Thus (2b) applies easily only to a few animals, like foraging vultures and sandpipers, and (2a) only to species like African hunting dogs, wolves, lions, group-fishing pelicans, and group-hunting fish. That leaves responses to predators as the probable evolutionary basis, or function, of all other actively formed groups, including most primate species and all of the great herds of ungulates and schools of fish.[9] Because laws are only made by humans living in the kinds of social groups we call societies, understanding group living is evidently closely related to understanding systems of laws.

Causes of Human Groupings

Everyone knows that early groups of humans are postulated to have been hunters of large game. Their predecessors almost certainly lived in groups for the reason that, probably like all modern group-living nonhuman primates, they were the hunted rather than the hunters. To all indications man is the only primate who became to some significant extent a group hunter—the only group-living primate who, at least for a time, escaped having his social organization essentially determined by large predators. In this light, it may not be so startling that dog and wolf packs and lion prides are social groups with which we empathize to a great degree—social groups that fascinate behaviorists because of parallels and complexities that are not clearly established elsewhere outside the human species. The human brand of sociality thus appears to be approached from two different directions—by various other primates because they are man's closest relatives, and by canines and cats because they most nearly do, socially, what humans did for some long time.

But the organization and maintenance of recent and large human social groups cannot be explained by a group-hunting hypothesis.[10] The reason is obvious: The upper size of a group in which each individual gained because of the group's ability to bring down large game would be rather small. As weapons and cooperative strategies improved, then, owing to the automatic expenses of group living, group sizes should have gone down. Instead they went up—right up to nations of hundreds of millions.

Human nations of millions of individuals, each potentially reproductive, appear to be unique in the history of the earth. There is no parallel, as often supposed in the past, with the social insects, in which one or a few females do all of the reproducing and the rest are closely related sterile workers and soldiers. Chimpanzees, baboons, and macaques are probably our closest counterparts in this regard, and their social organization likely never escaped strong effects from predators of other species (one of the most important of which may well have been our own ancestors).

What, then, did cause human groups to keep right on growing? If we hold to the arguments described above, what forces could possibly account for the rise of what anthropologists have called the "nation-state"? The uniqueness of human group sizes, as well as the uniqueness of humans, suggests that unique, truly remarkable causes may be involved.

One possibility is that the early benefits of group living (such as group hunting, cooperation in irrigation, and a host of others) were so powerful that they produced humans with such strong tendencies to group that they developed the huge modern nations of today as more or less incidental effects. This argument construes humans as being considerably less flexible in their behavior than I would like to allow. It says, in effect, that we are captives of our genetic history, and are such compulsive group livers that we pursue the habit relentlessly despite deleterious effects on ourselves and our children, and despite its hindering effects on the reproduction of ourselves and our close relatives. Perhaps such an argument does not seem too remote to those who have regarded reproduction as a triviality in human history, or to those who do not recognize the degree of opportunistic flexibility that typifies the human organism. Any such argument, however, seems entirely impotent to me. Moreover, should this be the real reason for human sociality, alternative hypotheses should not easily apply.

But there is an alternative hypothesis, one recently proposed by several different writers, and one which seems to me reasonable, appropriately unique, and clearly relevant to all efforts to understand, govern, and perpetuate ourselves.[11] I will call it the "Balance-of-Power Hypothesis." This hypothesis contends that at some early point in our history the actual function of human groups—their significance for their individual members—involved the competitive and predatory effects of other human groups and protection from them. The premise is that the necessary and sufficient forces to explain the maintenance of every kind and size of human group, extant today and throughout all but the earliest portions of human history, were (a) war, or intergroup competition and aggression, and (b) the maintenance of balances of power between such groups. I emphasize that this is hypothesis, not conclusion, and I state it in this simple radical form to make it maximally vulnerable to falsification.[12]

The model deriving from this argument would divide early human history into three periods of sociality, roughly as follows:

1. Small, polygynous, probably multi-male bands which stayed together for protection against large predators. (By polygyny is meant not necessarily the maintenance of

harems, but simply that fewer males than females were contributing genetically.)[13]

2. Small polygynous, multi-male bands which stayed together both for protection against large predators (probably through aggressive defense) and because of the ability to bring down large game (perhaps, at certain times, entirely because of one or the other of these reasons).

3. Increasingly larger polygynous, multi-male bands which stayed together largely or entirely because of the threat of other, similar, nearby groups of humans.

I suggest that expressions of human social organization today are derived from this sequence, with the relative importance of each stage in understanding sociality in modern humans dependent upon the duration of the stage, the intensity of selection during it, and where it occurs in the sequence. I also suppose that we have been in the third stage so long that the influences of the first two stages are relatively minor. The latter assumption departs dramatically from arguments of other writers, but that is not critical to the arguments that follow.

To relate the scenario I have just constructed to the thinking of archaeologists and anthropologists on the problem of the rise of nations, I would call attention to the recent review of Flannery (1972) on the evolution of civilizations, to Carneiro's (1970) paper, and to Webster's (1975) critique of Carneiro's argument.[14] Flannery describes the modern range of human societies from small hunting-gathering bands of fewer than two hundred individual affiliates through tribes and chiefdoms to huge industrial nations. He notes that all large nations had to pass through at least some of the smaller stages to reach their present condition, and shows that archaeological evidence regarding the earliest known dates for the three classes of societies larger than band societies suggests the appropriate chronology from small groups to large.[15] Then he asks what "prime mover" could account for the trend toward larger, more complex states and nations?

After reviewing many extrinsic possibilities and finding each either unnecessary or insufficient, Flannery seems to follow an approach frequently resorted to by biologists and social scientists alike in this kind of situation; he seems to seek the reasons for the nature of society in its internal workings. This approach leads to hypotheses like the one discarded earlier here—hypotheses of orthogenesis or genetic, physiological, or social "constraints" or "inertia." In biology it leads to what are termed arguments from physiological limitations—tendencies to explain each attribute as the maximum that could be achieved in a certain direction despite continued favoring of directional change. In effect, it requires that one explain ultimate causes by proximate causes rather than vice versa, and it is at best a vulnerable argument.[16]

Anyone who invokes proximate limitations to explain extant phenomena of life is in effect denying the power of the evolutionary process to produce some perceived or imagined effect. Sometimes there are valid arguments from adaptation for such explanations. An example is the argument that no more than two sexes exist because the presence of three or more would automatically cause an ecologically inferior sex to become less valuable as it became rare and eventually to disappear; with two sexes individuals of the rare sex automatically become more valuable.[17] But to deny adaptive significance because of supposed constraints on natural selection is perilously close to asserting either that marvels like humans and honeybees are impossible, or that they are entirely predictable. Moreover, the rapid directional changes induced by human selection, especially upon domestic animals and plants, and the diversity of effects achieved within species by different directions of selection in only a few generations, tend to deny any long-term significance of genetic constraints and attest to the potency of selection.

How, though, does Flannery dismiss the hypothesis suggested here, that of intergroup competition? He notes that intergroup aggression has evidently been continuous throughout history in many parts of the world where large nations have never evolved. Like Webster and others, he concludes from this that, while war may be necessary to explain the origin of the state, it is not sufficient.

But these authors do not explicitly consider the question of balances of power. Balances of power depend to some extent on physiographic and other extrinsic environmental circumstances, and they may as well exist between tiny New Guinea tribes as between nuclear powers.[18] Moreover, aspects of intergroup conflict among such people, which are commonly referred to as ceremonial or ritualistic, may actually reflect the importance of balances of power; examples are elaborate bluffing and the intensity of concern with avenging each death. Balances of power are also significant within groups, continually denying to individuals and subgroups the possibility of initiating individualistic reproductive strategies or fragmenting the larger group by secession or fission.

If, for whatever reasons, growing imbalance through one-sided expansions of some groups, or superiority in weapons or some other regard is not possible, then large nations may never appear. One test of a balance-of-power hypothesis would involve checking to see if physiographic or other barriers reduce the effectiveness of coalitions or the likelihood of unity across areas of increasing size, preserving the balance at low group sizes (Carneiro's "environmental circumscription"). Another test would be to see if empires have tended to develop in pairs or groups, or centrally nested inside multiple smaller competitors, and to disintegrate when they lacked suitable adversaries (Carneiro's "social conscription"). Even a very general knowledge of history suggests that these things have been true. Carneiro's hypothesis, and the analyses of Flannery and others, seem to me to put us on the brink of modifying to acceptability a hypothesis of just the sort discussed here (and in somewhat different forms by others).

Across the past several decades, failures in the social sciences to locate broad explanatory generalizations have led to the tendency to suppose that, since singular explanations have been singularly absent, it is more appropriate to seek or rely upon multiple causations than to accept singular ones—even, it seems, if one of the latter should appear sufficient! Perhaps it is vile and degrading to expect that singular explanations can be derived at any level for the complex phenomena of human behavior, but not nearly so much so as to deny them on that basis alone. The new evolutionary arguments about group living summarized above, for example, simultaneously cast doubt on the older "group-function" explanations and imply that a singular basis is both possible and likely.

Now let me review the steps by which I arrived at the hypothesis that the rise of the nation state depended on intergroup competition and aggression, and the maintenance of balances of power with increasing sizes of human groups. First, Williams's convincing argument that selection usually is effective only at individual or genic levels forced a search for reasons for group living that would balance its automatic costs to individuals. The available reasons have proved to be small in number, and only one, predator protection, appears applicable to large groups of organisms, including those of humans. For humans a principal "predator" is clearly other groups of humans, and it appears that no other species or set of species could possibly fulfill the function of forcing the ever-larger groups that have developed during human history. Carneiro[14] and Flannery[14] essentially eliminated as "prime movers" all of the other forces previously proposed to explain the rise of nations, and I think their arguments are reasonable. Flannery and Webster also eliminated intergroup competition as a prime or singular force, and they sought causes of the rise of nations within societal structure. This last procedure is here deemed unsatisfactory because it leads to the explanation of ultimate factors by proximate mechanisms rather than vice versa. Moreover, Flannery's rejection of intergroup aggression as necessary but not sufficient is deemed inadequate because he did not specifically consider intergroup aggression in terms of the maintenance of balances of power. His elimination of other factors may or may not be satisfactory; the fact is, in light of the realization that the automatic expenses to individuals which accompany group living are generally exacerbated as group sizes increase, none of the supposed causes for the rise of nations except balances of power seems even remotely appropriate; nor do they serve any better when grouped and regarded as multiply contributory.

The kind of argument I am making here cannot fail to be disconcerting, or even bizarre, to many modern scientists, philosophers, and humanists, especially those who are reasonably well satisfied with their present way of looking at things; it is too novel for anything else to be the case. There is only one acceptable basis for rejecting it, however, and that is by demonstrating its weaknesses or error. I believe that application of Williams's refinement of Darwinism, as I am attempting to do here, seriously threatens current philosophical thinking at its base and cannot fail to alter dramatically the theoretical underpinning of the social sciences, not in the sense often imputed to Wilson (1975)—that the formulation of normative ethics must depend upon and derive from biology—but in the sense that interpretations of history, and predictions about the future of humans not yet cognizant of these matters, will be facilitated by this kind of thinking more than by any other aspect of human understanding. Of course, if I am right, how any desired social, ethical, moral, philosophical, or legal situation is realized will also be massively affected by the analytical process proposed by evolutionary biology. My view of the potential flexibility of human behavior, however, causes me to deny any other necessary effects of knowledge of evolution upon the efforts of human beings to manipulate their future (see below).

Group Living and Rules

With these arguments and hypotheses about the history of human society, we may be in a position to develop a clearer overall view of the backgrounds and significance of societal rules and standards.

In the first place, we are led to hypothesize that rules in some fashion represent the wishes of individuals and relate to reproductive competition among individuals within groups, with the additional constraint that individual reproductive success within groups depends to some extent on the success or maintenance of the group as a whole. In other words, we might hypothesize that individuals behaving consistently in respect to the long evolutionary history of humans should work to preserve their group and keep it healthy while simultaneously striving, as far as possible, to convert their groups into clans of descendants and other genetic relatives related as closely as possible to themselves.

It is relevant that efforts to cause changes in the behavior of populations only work when the individuals in the population regard them as personally advantageous: It has to be to the *individual*'s advantage to reduce family size, conserve fuel, or treat his neighbor right; or it has to be to his disadvantage not to do so. Cooperative subgroups, like corporations, are not likely to follow courses that match the interests of the whole group, as in avoiding pollution, resource depletion, or profiteering, unless (a) the penalties imposed by the whole group are sufficient to eliminate the profit in selfish behavior, or (b) a threat external to the entire group makes it temporarily profitable to direct efforts primarily to sustaining it (not the least reason for which is the "public relations" effect from altruism or heroism in such situations).[19]

It is the purpose of laws to cause these things to be true, and to be regarded by individuals as having this function; individual members of a group tend to obey the laws and work for the common good with the least encouragement at times when a group is obviously threatened by external forces—indeed, these are the times when even huge nations form alliances with one another.

What Is Justice?

To relate these musings by a biologist to the thoughts of legal philosophers and sociologists, we can turn briefly to perhaps the most widely asked question about societal laws, the very prominence of which supports the individualistic interpretation of history being defended here. The question is, "What is justice?" I will start with an essay on that topic by Hans Kelsen, which opens his book of the same title.[20] Kelsen notes that:

> No other question has been discussed so passionately; no other question has caused so much precious blood and so many bitter tears to be shed; no other question has been the object of so much intensive thinking by the most illustrious thinkers from Plato to Kant; and yet, this question is today as unanswered as it ever was. It seems that it is one of those questions to which the resigned wisdom applies that man cannot find a definitive answer, but can only try to improve the question.

Kelsen goes on to define justice as "social happiness," and he says:

> It is obvious that there can be no "just" order, that is one affording happiness to everyone, as long as one defines the concept of happiness in its original, narrow sense of individual happiness, meaning by a man's happiness, what he himself considers it to be. For it is then inevitable that the happiness of one individual will, at some time, be directly in conflict with that of another.

> Where there is no conflict of interests, there is no need for justice. A conflict of interests exists when one interest can be satisfied only at the expense of the other; or, what amounts to the same, when there is a conflict between two values, and when it is not

possible to realize both at the same time; when the one can be realized only if the other is neglected; when it is necessary to prefer the realization of the one to that of the other; to decide which one is more important, or in other terms to decide which is the higher value, and finally: which is the highest value.

Kelsen notes that justice must be relative and incomplete, and can only be regarded as ideal or absolute if it is accepted (by everyone!) as having been determined by an ideal or absolute being, such as God—a view expressed by a long succession of philosophers.

Justice is necessarily incomplete, and laws are fluid, then, because people strive. To understand sociality and the sociology of law it would seem useful to know what people are striving for. This too is an old question. Kelsen, and the American Bill of Rights, say they are striving for happiness, and this is no doubt true. Happiness, though, as the different versions of a prevalent adage tell us, is many things. It is eating and sex and parenting and warmth and touching and ownership and giving and receiving and loving and being loved; it is eating when hungry, drinking when thirsty, coolness when it's hot, and warmth when it's cold; it is the cessation of pain and the onset of pleasure; it's finding a way to win; it's having a magnificent idea or a grandchild.

In some sense these are all biological things. There is probably nothing on the list that doesn't bring a happiness equivalent to at least one nonhuman organism as well as to humans.

Biologists ask another kind of question about the organisms they study: *Why* does that particular event or stimulus bring pleasure or happiness? As a result, biologists are able to generalize about happiness and see it eventually as an evolved means to an end. The neural connections that cause a bare foot to hurt when it comes down on a sharp object are not there accidentally; nor are those that cause a pleasurable sensation when a ripe fruit is placed inside the mouth. These are evolved correlations. Pleasure and happiness associate with events and stimuli that are beneficial to us in the usual environments of history.

Why should it be so? It should be so only if "beneficial" is defined, in terms of history, as leading to reproduction, i.e., as leading to genetic survival. Whatever, in the past, led to increased reproduction was likely to save itself, to cause its own perpetuation. Whatever did not was at least irrelevant, and, as such, deleterious to its own survival as an alternative to anything that correlated with reproduction. In this light we can understand why intense pleasure should be associated with the opportunity to benefit our offspring—or improve their situations by personal sacrifices—even, if necessary, by giving our lives for them in excruciatingly painful fashion.

Biologists divide the lives of organisms into two stages: resource-garnering (growing, maturing, becoming wealthy or powerful or clever) and resource-redistribution (reproduction, using power, wealth, and wisdom to produce and assist descendants and other relatives).

Now we can suggest that what humans have evolved to strive for is to reproduce and to reproduce maximally—indeed, to *out*-reproduce others.[21] Happiness, then, is an end for the individual only in the sense that it is achieved by acts leading to reproduction. Happiness is a means to reproduction (That this is strictly true only in historical terms and that happiness can obviously be diverted from the goal of reproduction can be ignored for the moment.)

In other words, the striving of organisms can be generalized on solid grounds, and it is not hedonistic at all but reproductive; in historical terms hedonism is itself reproductive, and when it is not we expect it eventually to be abandoned.

I believe that these thoughts give us a way of understanding why human striving is incompatible with the concept of ideal, pure, or complete justice. First, humans strive as individuals or subgroups rather than united wholes. Second, there is no automatic finiteness to their striving because success can be measured in no way except in relation to one another: It follows that their separate strivings conflict, and sometimes involve direct thwarting of one another's efforts. Finally, they are continually altering their strivings to increase their success in the changing situations of life, and thereby introducing additional changes.

The differences of interest that legal philosophers discuss are thus based on differences of reproductive interest, and ultimately on genetic differences, and they are not likely to be resolved, in any absolute sense, by allowing given amounts of reward, payment, or returns on investments.

Our interests, then, are turned to several items, the first of which may be degrees of genetic relatedness or overlap among interacting individuals. We can even suggest that it is no accident that Kelsen's examples of the complexities of justice are: (a) two men in love with the same woman; (b) King Solomon's threat to divide a disputed child between the two women who claimed it, with the intent of giving it to the woman who loved it too much to allow it to be hurt; and (c) two men in competition for the same prestigious job.

The basis for conflicts of interest among individuals—hence, the basis for the unresolvability of the question of justice—evidently derives from our history of reproductive competiton operating primarily at the individual level. I am saying bluntly that social conflict derives from biological facts. To me this suggests that our best chance for diminishing social conflict lies in better understanding of its biological basis.

Since most people today live in nation-states, we must also be interested in the nature of such societies and the possibility of generalizing the basis for the systems of law by which nation-states manage to function. Stein and Shand, in their 1974 book, *Legal Values in Western Society,* argue that order is "the primary value with which law is associated." But they answer the question, "What is law for?" by saying that the "three basic values of the legal system" are "order, justice, and personal freedom." From the arguments I have just made, and those that follow, I suggest instead that law is "for" but one thing: the preservation of order; and that justice and personal freedom, to whatever extent they are sought or approached, are also for the purpose of maintaining order. Order is valuable to everyone if extrinsic threats to the group are sufficiently severe, and the group is of no value if there are no such threats. In times of little or no extrinsic threat, on the other hand, laws are most valuable to those who lopsidedly control resources. These people generally include the wealthy and secure (versus the poor and insecure), parents (versus individual offspring), and older people (versus younger people). Revolutionaries (those willing to destroy order) must either (a) perceive themselves to be in a very bad position within society; (b) suppose that no significant threats to the group exist at the moment (so that internal dissension would not lead to worse troubles for themselves as a result of outside forces); or (c) have support from outside the group that seems to them to promise a better situation as a result of destruction of the existing order, and perhaps even their group as constituted.

And so we are returned to the biologist's view of organisms' lives being divisible into the activities of resource-garnering and resource-redistribution. I think we are talking about the basis for everyday phenomena such as the so-called "generation gap," inheritance laws, changes in occupation at midlife, racism, and reasons for racism's effects falling more heavily upon one sex (male) of the minority group than upon the other.

What I am saying, in many parts of this essay, I recognize to be essentially common sense. It seems to me that this is true to the extent that appropriate explanations are being approached. On the other hand, even common sense should become infinitely more sensible in the context of a history of differential reproduction, and I want to pursue that possibility further.

Reproductive Competition and Law-breaking

I have argued that the *function of laws is to regulate and render finite the reproductive strivings of individuals and subgroups within societies, in the interest of preserving unity in the larger group* (all of "society" or the nation-state). Presumably, unity in the larger group feeds back beneficial effects on those segments or units which propose, maintain, adjust, and enforce the laws. Partly because of continual shifts of interests, changing coalitions, and power adjustments, it is not likely always to do so evenly, or in such fashion as to cause all individuals to benefit equally from group unity; hence, the value of "federal" government.

As a preliminary test of the general model that laws function to place limits on reproductive striving,[22] several obvious predictions may be considered. First, laws should be constructed so as to regulate competitive striving, and the severity of punishment should reflect the severity of deleterious effects on the reproduction of others. Capital punishment is generally correlated with murder, which destroys the victim's ability to reproduce; treason, which potentially lowers, at least slightly and perhaps massively, the reproductive fitnesses of everyone else in one's society; rape, which may directly interfere with a man's chances of reproducing via his spouse, sister, daughter, or other female relatives. Rape laws are particularly interesting to consider, since, because it is a nonfatal kind of assault, rape may not at first seem to be an appropriate transgression for the imposition of capital punishment.

If we were to select a category of striving that is centrally important, restricted to a definite part of the human life span, and more intense in one sex than the other, then we should be able to plot changes in the intensity of that striving—say, by age—against changes in the likelihood or rate of lawbreaking.

Suppose we choose sexual competition, or competition for mates, including all of the various activities involved in increasing one's desirability as a mate, hence, one's ability to select among a wider array of potential mates. First, sexual competition is demonstrably more intense among males than among females; and one can easily show from apparently accumulated differences between modern human males and females powerful evidence that this has consistently been the case during human history, and that a general consequence is that the entire life-history strategy of males is a higher-risk, higher-stakes adventure than that of females.[23] This finding leads to the prediction that lawbreaking will occur more frequently among males, which of course is already well known.[24] It also seems to predict that laws are chiefly made by men (as opposed to women) to control men (as opposed to women). That laws are made by men to control men is suggested over and over again in the structure and application of laws. As perhaps the prime example, it has become painfully clear because of attention recently brought to bear by women's groups on the application of laws against rape that the female victims are treated like pawns by the collections of (mostly) males who enforce the laws, judge and punish the offenders, and are indirectly wronged because of affinal or kin relationship to the victim. One might say that rape victims appear to be treated as if rape laws were designed to protect them only in the sense that rape wrongs the males to whom they "belong" or might have belonged and reduces their value or attractiveness to those males; and in the sense that, when rapists are free to act, the interests of all *men* are in jeopardy. It appears that the female victims of rape may be only incidental to the development and application of rape laws. Under current circumstances, the most pathetic of all rape victims is probably the female (1) without a male who has at that moment a proprietary interest in her welfare, such as a father, brother, husband, or sweetheart; (2) whose male defenders are already somewhat tentative in their allegiance to her; (3) who was raped by such a person or in such a circumstance as to pose little threat to other males in society; or (4) was raped in a fashion or circumstance that reflects particularly severely on her future desirability as a mate. It is relevant that in most states a man cannot be accused of raping his wife, and, until very recently, a man could kill his wife or her lover without being accused of murder while a wife could not.[25]

Lawbreaking is also expected to be concentrated at those periods in life, or those ages, when competitive striving is most intense or most crucial. Competition for mates is greatest just before the usual ages of marriage, and the extent to which an individual is able to begin effectively to climb the ladder of affluence may also be determined at about the same ages. Lawbreaking is strongly concentrated during ages 17-22 in technological nations.[26] These are also the precise ages at which Yanomamö Indians, and probably men in most societies, suffer their highest mortality, mostly from intergroup aggression, and also the ages of highest likelihood of military induction.[27] These ages immediately precede those at which marriage is most frequent.

Lawbreaking is expected to be higher in individuals or groups most inhibited from climbing the ladder of affluence or using the system legally to accumulate resources. Moreover, lawbreaking should be even more heavily concentrated in males who are more or less publicly identified as likely to have such difficulties. Thus:

(a) Lawbreaking should be, and is, higher in minority-group males than in majority-group males.[28]

(b) Lawbreaking should be, and is, higher in publicly (e.g., physically) identifiable minority-group males than in those not publicly identifiable.[28]

(c) In the absence of publicly identifiable minority groups, lawbreaking should be highest in young males whose families give them least assistance in climbing the ladder of affluence. The most dramatic correlation found by Ferracuti and Dinitz (1974) with delinquency in a racially homogenous Puerto Rican situation was with lowering social status of the boy's family.[29]

Lawbreaking is also higher in families lacking one parent—especially the father—in families that are less religious, and in families with less control over their children and who give their children less assistance, encouragement, and attention, but who punish them more often.[29]

Finally, we may consider alternative strategies of reproductive competition among males, their distribution, and their consequences for patterns of lawlessness. I assume that freedom of opportunity to "climb the ladder of affluence" is a crucial aspect of sexual competition which, when available, will supplant all others—that is, that males who are either affluent or have a very high likelihood of becoming affluent are among the most desirable mates. Such males are unlikely to be lawbreakers, except in the context of using the system to further their affluence illegally, such as by income tax tricks or misuse of power associated with affluence.

In contrast, one expects alternative strategies, such as behavior that can be considered under the general label of "machismo," or flash and braggadocio, to be concentrated in individuals or groups whose likelihood of climbing the ladder of affluence ("using the system" effectively) is lowest, and especially when this low likelihood is publicly projected by inescapable identification with a disadvantaged group, such as a minority that the system discriminates against. "Macho" strategies of sexual competition are at once declarations of desirable qualities other than affluence, denials of the value of affluence as usually measured, rejections of the system, and declarations of a degree of disdain and independence with regard to the rules of the system. Sexual competition by macho behavior is almost by definition a declaration of lawlessness, or willingness to break the law.

In summary, the predictions may be met that (a) macho behavior, such as flashy dressing and abandonment of spouses and families, and (b) lawbreaking will be concentrated in men who (i) are young, (ii) lack wealthy, influential, successful, or powerful relatives, and (iii) are recognizable as members of minority or other disadvantaged groups. It is an obvious corollary that somehow to equalize the possibilities of individual men of all classes and origins of "climbing the ladder of affluence"—at least in terms of personal capabilities (and how they are seen by their possessors)—would provide the most reliable if not the easiest way of reducing the problem of overcrowding in prisons. Perhaps not surprisingly, this conclusion is consistent with those arrived at from entirely different approaches, and after very extensive examinations of correlates of crime and delinquency.[30]

Nepotism and Reciprocity

At this point I think it is necessary to distinguish explicitly two general classes of social interactions which have different characteristics and different outcomes. Although their differences and similarities are thoroughly explored elsewhere,[31] their changing relationship to societal laws in different kinds and sizes of groups requires a brief summary here.

Nepotism implies that benefits are given by one individual to another without reciprocation, the gain to the first individual resulting from the genetic overlap between the two. Reciprocity implies that benefits are only given when there is a high likelihood of a compensating return to the phenotype of the benefit giver (but, conceivably, the return could be to the phenotypes of relatives of the benefit giver). Only reciprocity can evolve within groups of unrelated individuals. In groups of equally related individuals (or individuals who cannot respond to differences in degrees of relatedness and can only evolve to respond in terms of average relatedness), nepotism will also evolve such that only two kinds of individuals will be distinguished: group members and nongroup members. Reciprocity can also evolve in such groups.

What happens in groups of variously related individuals, within which the individuals can respond to the differences in relationships? Obviously, when reciprocity is unlikely, closer relatives should be favored in beneficence. But so should they if there is any doubt about reciprocity, as there must always be. One can better afford to lose, and less afford to cheat, in reciprocity with a relative, and more so with a closer relative than with a more distant one.

Even parental behavior is, like reciprocity, an investment involving the risk that it may not yield a suitable return. If acts of nepotism should be directed at one's closest relatives, among those with equal needs, so should acts of reciprocity, assuming no variation in cost-benefit ratios. This means that in groups of variously related individuals, nepotism and reciprocity will always tend to be intricated. Thus, reciprocity can evolve alone, but only in groups of unrelated individuals; but nepotism evidently cannot evolve independently of reciprocity because reciprocity will always be potentially a factor in the social interactions of groups within which nepotism can evolve. This fact may be responsible for many of the confusing aspects of human systems of kinship cooperation and altruism that have led social scientists to doubt their derivation from a history of differential reproduction.[32]

Elsewhere I have argued that nepotistic behavior toward non-descendant relatives evolves out of parent-offspring interactions, and that reciprocity derives from nepotism.[33] Here I suggest that authority, in regard to regulation of social interactions, originates in parental authority, and that parental control of resources is a major aspect of rule construction and enforcement. From this beginning the extension to nepotism is small in the forms of assisting relatives in efforts to obtain repayment of debts owed them, avenging wrongs done to relatives, and accepting responsibility for debts incurred by relatives.

The development of nation-states correlates with the suppression of nepotism, a rise of concern with law and order, a rise of belief in the authority of divine beings or rulers, and an acceptance of nobility and divinity in leaders and rulers. I ask whether there is a connection between these changes and the original parental authority, with the origins of divinity being reverence toward deceased powerful ancestors and the effort to use their presumed wishes and authority to promote order, and to use the ability to convince others of either special knowledge of such ancestors' wishes or communication with them to succeed to leadership and power (hence, I suppose, unusual reproductive opportunities). I do not suppose it an accident that God should have come to be regarded as a "Father in Heaven."

Changes in Rules with Development of Nation-States

I have already argued that the rise of nation-states occurred as a result of the interactions of neighboring competitive and hostile groups expanding their alliances and cementing unities in a balance-of-power race. Now I suggest that rather than the rise of the nation-state being understandable through knowledge of its internal workings,[34] the internal workings of the nation-state are understandable only in terms of the reasons for its appearance, namely, intergroup aggression and competition. Let us examine briefly the sources and kinds of rules in the different kinds of societies compared by Flannery and widely regarded as representing stages preceding the nation-state. They are bands, tribes, chiefdoms, and stratified societies. In drawing heavily from Flannery's review, which I find consistent with other writings on this topic, I note that Flannery wrote in total unawareness of the arguments I am making.

Flannery notes that the only "segments" of *bands* are "families or groups of related families" and their "means of integration are usually limited to familial bonds of kinship and marriage, plus common residence. Leadership is informal and ephemeral; division of labor is along the lines of age and sex; and concepts of territoriality, descent, or lineage are weakly developed."

In extant band societies there is little heritable wealth. Social interactions are said to be based largely on "reciprocity," but the term has been used by anthropologists who did not distinguish nepotism and reciprocity. Wiessner[35] has evidence that in Kalahari Bushmen "reciprocity" is essentially limited to known genetic relatives, and it is practiced more with closer relatives than with more distant ones.

What authority there is in band societies seems to derive largely from parents and collections of parents, especially older men. Relatives defend and avenge one another, and they are expected to do so. The social "cement" of band societies is clearly nepotism.

Tribes are larger groups "whose segments are groups of families related by common descent or by membership in a variety of kinbased groups (clans, lineages, descent lines, kindreds, etc.). . . . Ancestors are often revered, and it is believed that they continue to take part in the activities of the lineage even after death. . . . Since 'tribes,' like bands, have weak and ephemeral leadership, they are further integrated (and even, it has been argued, regulate their environmental and interpersonal relations) by elaborate ceremonies and rituals. Some of these are conducted by formal 'sodalities' or 'fraternal orders' in which members of many lineages participate. . . . 'Tribes' frequently have ceremonies which are regularly scheduled . . . [and] may help to maintain undegraded environments, limit intergroup raiding, adjust man-land ratios, facilitate trade, redistribute natural resources, and 'level' any differences in wealth which threaten societies' egalitarian structure. . . ."[36]

It seems to me that Flannery may be describing the rudiments of laws that hold together groups of not-so-close relatives by imposing and maintaining restrictions on reproductive competition. He describes from archaeological finds "pottery masks . . . countless figurines of dancers . . . incredible accumulations of shell rattles, deer scapula rasps, turtle shell drums, conch shell trumpets . . ." which suggest not only ceremony but significant differentials in heritable wealth.

Chagnon[37] has noted that Yanomamö Indians may not mention the dead. Yet the Yanomamö, he assures me, otherwise fit Flannery's usage of tribes. This sensitivity around the use of ancestors may indicate the difference between allowing and not allowing succession to power and influence by identification with powerful deceased relatives. The Yanomamö tend to fission when a powerful ancestor dies, with the sizes of groups at fissioning correlated with degree of genetic relatedness in the groups. Tribes discussed by Flannery revere such ancestors, and, one supposes, may use them to enhance unity in their societies of groups of related families. The power of parents and the unity of nepotism thus still appear the major source of authority and rules in the tribal societies to which Flannery refers.

Chiefdoms are still larger groups in which "lineages are 'ranked' with regard to each other, and men from birth are of 'chiefly' or 'commoner descent.' " Such chiefs "are not merely of noble birth, but usually divine; they have special relationships with the gods which are denied commoners and which legitimize their right to demand community support and tribute . . . the chief . . . may be a priest . . . the office of 'chief' exists apart from the man who occupies it, and on his death the office must be filled by one of equally noble descent; some chiefdoms, maintained elaborate genealogies to establish this . . ."

"Since lineages are also property-holding units, it is not surprising to find that in some chiefdoms the best agricultural land or the best fishing localities are 'owned' by the highest-ranking lineages . . . high-ranking members of chiefdoms reinforce their status with sumptuary goods, some of which archaeologists later recover in the form of 'art works' in jade, turquoise, alabaster, gold, lapis lazuli, and so on."[38]

In chiefdoms it would appear that sources of authority have become more significant than in small groups, sometimes shifting from parental authority to deceased ancestors to gods representing extensions of such deceased ancestors. One also notices that the "office" of chief has itself become a vehicle of potential reproductive success for the individual who attains it—hence, itself a sought-after position (i.e., it is no longer strictly a vehicle for nepotism to the entire subject group, as in the case of a family patriarch). This opportunity is accepted and allowed by group members, perhaps because of the value to them of competition for the position of chief, which increases the likelihood that their leader will be a capable one.

Finally, "The state is a type of very strong, usually highly centralized government, with a professional ruling class, largely divorced from the bonds of kinship which characterize simpler societies. It is highly stratified and extremely diversified internally, with residential patterns often based on occupational specialization rather than blood or affinal relationship. The state attempts to maintain a monopoly of force, and is characterized by true law; almost any crime may be considered a crime against the state, in which case punishment is meted out by the state according to codified procedures, rather than being the responsibility of the offended party or his kin, as in simpler societies. While individual citizens must forego violence the state can wage war; it can also draft soldiers, levy taxes, and exact tribute."[39]

Nepotism displays a peculiarly altered condition within the nation-state, as compared to the smaller kinds of human societies in which it may represent the basic social cement. Nepotism obviously cannot be the social cement of nation-states of millions or hundreds of millions of individuals; only reciprocity can fulfill this function; and, of course, the interactions of individuals in nation-states are always organized around barter, currency, and various kinds of legal obligations and documents which ensure that debts are paid. So, within large nation-states we retain ties chiefly to the immediate family, and we tend to identify as immediate family parents, offspring, and siblings. Because there is a correlation between the uniquely human phenomenon of socially imposed monogamy and the nation-state, everyone in the immediate family is related to Ego by 1/2; nephews, nieces, aunts, and uncles by 1/4; first cousins by 1/8. More distant relatives are generally classed just that way—as "distant relatives." We do not usually organize into clans, and when we do we are usually regarded as behaving outside the law. So nepotism in the nation-state seems focused on the individual and the immediate family, with its vestiges outside the family likely often "misdirected," in historical, reproductive terms (especially in modern societies with the high and novel degree of mobility of our own) to neighbors, roommates, or others whose social relationships to us mimic those of relatives in the past. The functional relationships between nepotism and reciprocity described earlier thus correspond to the roles and relative prominences of these two related aspects of sociality in the different kinds and sizes of human social systems extant today and believed roughly to correspond to stages in the development of the nation-state.[40]

It seems to me that the categories into which the laws of nation-states can be arranged are commensurate with the biological arguments made above: (1) those which prevent individuals or groups from too severely interfering with the reproductive success of others, (2) those which prevent individuals or groups from too dramatically enhancing their own reproductive success, and (3) those which promote industry and creativity in individuals and groups in ways that may be exploited or plagiarized by the larger collective. Examples of laws in these three categories, respectively, are those concerning:

a. murder, assault, rape, kidnapping, treason, theft, extortion, breach of contract.
b. polygamy, nepotism, tax evasion, draft evasion, monopolies.
c. patents, copyrights, wills.

As a final comment, I cannot resist noting that the Ten Commandments look like a legal prescription for the maintenance of a nation-state. I find it easy to interpret the first four, all of which deal with paying homage to God and not breaking his laws, as referring to the importance of preserving the large group. I am impressed that 40 percent of the rules seem concerned with this issue.

The fifth says that we should also honor our parents. This is commensurate with arguments advanced so far: Half of the commandments thus deal with respecting sources of authority or not tampering with current distributions of resource control. These are the commandments that include threats of retribution. Even the fifth concludes its admonishment with the phrase "that thy days may be long in the land which the Lord thy God giveth thee," probably, however, referring to the family's survival rather than that of the individual—hence, effectively referring to genes rather than individuals!

The next four commandments tell us not to kill, commit adultery, steal, or lie. The tenth tells us not even to think about it. I am particularly impressed with the tenth commandment because, in my experience, humans tend to regard as first novel and bizarre, then ludicrous and outrageous, the suggestion that their evolutionary history may have primed them to be wholly concerned with genetic reproduction (in the environments of history). How does it happen, that in the course of evolving consciousness as a state into or through which *some* of our behaviors are expressed, we are so emphatic (and public) about rejecting this seemingly ever-so-reasonable one? Simultaneously we seem to reject the possibility that what we are truly about could be something we hadn't really thought of—personally and individually. It makes one wonder, quite seriously, if there might not be something incompatible about telling young children all about natural selection and rearing them to be properly and effectively social in the ways that we always have.[41]

Evolution and Normative Ethics

The arguments given above, and the cited references, make it clear why I believe that evolution has more to say about why people do what they do than any other theory. In contrast, my answer to the question: "What does evolution have to say about normative ethics, or what people *ought* to be doing?" is: "Nothing whatsoever." Apparently this response is so startling that I am required to explain it.

I have two reasons for giving this answer. The first is that I regard humans as sufficiently plastic in their behavior to accomplish almost *whatever they wish*. The emphasis is on the final phrase because this is the crucial question.

There is an unfortunately prevalent attitude that to suppose an evolutionary background for behavior automatically supposes a predictable future into which we are helplessly cast as a consequence of the ontogenetic determinism produced in us by the history of selective action on our genes. The feeling seems to be that all evolution has to offer is information about our inevitable route through history. No one wants to know all about his future, unless the knowledge, paradoxically, promises to help him change it; and most people doubt anyway that such knowledge is possible. I am sure these feelings give rise to one kind of anti-evolutionism.

People who think this way are missing the fact that the life histories of individual organisms and the fates of species are predictable, in evolutionary terms, only to the extent that environments and their effects are predictable. For a species whose individual members possess cognitive and reflective ability, and the power of conscious prediction and testing of predictions, even the knowledge of its evolutionary history, and the interpretation of its individual tendencies in different ontogenetic environments on account of that history, become parts of the environment that determine its future. Indeed, I am contributing to this book solely because it seems to me that no other aspect of the human experience could possibly be so massively influential upon our future as a clear comprehension of the reasons, and therefore to some extent the nature, of the fine tuning of our personalities, individually and collectively, from the effects of an inexorable process of differential reproduction during our history.

I am saying that what a knowledge of evolution really offers us, in terms of the future, is an elaboration not a restricting of ontogenetic possibilities, of life history or life-style opportunities, and of collective potential for accomplishing whatever may be desired. It does this by telling us who we really are, and, therefore, how to become whatever we may want to become. Evolutionary understanding, then, more than anything, has the power to make humans sufficiently plastic to accomplish *whatever they wish*. This grandiose notion, of course, loses all its glamour if there is any doubt at all about the centrality of evolutionary theory as explanatory of human nature.

My second reason for denying that evolutionary understanding carries lessons about what we *ought* to be doing involves the background of such notions. Ethical structures have been developed throughout history without any extensive direct knowledge or conscious perception of the evolutionary process. If they have in any sense converged upon what might have generated in the presence of such understanding it has to be because individuals and collectives of individuals have identified rights and wrongs in terms of effects, ultimately, upon reproductive success. I have already argued that they have done so, and I think it is obvious that they have usually done so without any conscious knowledge of the relationship of reproductive success to either history or proximate rewards like sensations of pleasure or well being.

Does this mean, however, that opportunities for reproductive success necessarily must lie at the heart of our considerations of normative ethics for the future? I can see no reason for such an assumption.

So we are returned to proximate rewards, which have formed the basis for all systems of normative ethics anyway, without any particular evidence of their connections to ultimate reproductive success. No one needs evolutionary theory to identify proximate rewards in his own life, although such theory may clarify their significance to us. Moreover, anyone who rejects as a proximate reward to himself whatever may be identified as such from evolutionary considerations by definition cannot, in my opinion, be wrong.

However proper systems of normative ethics are identified, then, evolutionary considerations almost surely can help to achieve the goal. It must be obvious that I think that it can do this better than any other kind of knowledge. But evolutionary understanding has little or nothing to tell us about how to identify the goal. At most it may suggest that this question is destined to remain much more complex than we would like, that answers to it will change rather than become simple and static, and that it will never be answerable for all time at any particular time.[42]

NOTES

1. This step has been possible so far with very few traits of very few organisms, and only then on short-term bases or with rather low

levels of certainty. The truth is that we do not yet know with much confidence such things as why the dinosaurs became extinct. Attempts at syntheses of the sort implied here are likely, however, to be prominent features of evolutionary investigations in the future, and I think we may expect them to occur first in three areas: (1) the evolution of sterile castes in insects (because the underlying genetics in Hymenoptera—the major group involved—are asymmetrical owing to haplodiploid sex determination, and we understand them well); (2) the evolution of mating behavior in arthropods and vertebrates (because so much comparative analysis is possible and so much relevant information is available from related studies, like the use of genitalic morphology by taxonomists); and (3) the evolution of human social behavior (because we are so fascinated by it, and because paleontological and archaeological data continue to be gathered so rapidly).

2. The reasons for this assumption are not widely discussed. Some of them are the following: (1) altering directions of selection alters directions of genetic change in organisms; (2) the causes of mutations (chiefly radiation) and the causes of selection (Darwin's "hostile forces" of food shortages, climate, weather, predators, parasites, and diseases) are independent of one another; (3) only the causes of selection remain consistently directional for relatively long periods (thus could explain long-term directional changes); and (4) predictions based on the assumption that adaptiveness depends solely upon selection work. A prime example of the last is the history of sex ratio selection, traceable from the work of R. A. Fisher, *The Genetical Theory of Natural Selection*, New York: Dover, 1958; (1st ed. 1930), pp. 158-62; W. D. Hamilton, "Extraordinary Sex Ratios," *Science* 154 (1967): 477-88; R. L. Trivers and D. Willard, "Natural Selection of Parental Ability to Vary the Sex Ratio of Offspring," *Science* 179 (1973): 90-2; R. L. Trivers and H. Hare, "Haplodiploidy and the Evolution of the Social Insects," *Science* 191 (1976): 249-63; R. D. Alexander and P. W. Sherman, "Local Mate Competition and Parental Investment in Social Insects," *Science* 196 (1977): 494-500; R. D. Alexander, J. L. Hoogland, R. D. Howard, K. M. Noonan, and P. W. Sherman, "Sexual Dimorphisms and Breeding Systems in Pinnipeds, Ungulates, Primates, and Humans," to appear in *Evolutionary Biology and Human Social Behavior: An Anthropological Perspective*, ed. N. A. Chagnon and W. G. Irons, (North Scituate, Mass.: Duxbury Press).

3. See also R. C. Lewontin, "The Units of Selection," *Annual Review of Ecology and Systematics*, 1 (1970): 1-18; R. D. Alexander and

G. Borgia, "Group selection, altruism, and the hierarchical organization of life," *Annual Review of Ecology and Systematics* 9 (1978).

4. See references in the following: R. D. Alexander, "The Search for an Evolutionary Philosophy of Man," *Proceedings of the Royal Society of Victoria, 84 (1971): 99-120;* "The Evolution of Social Behavior," *Annual Review of Ecology and Systematics,* 5 (1974): 325-83; "The Search for a General Theory of Behavior," *Behavioral Science,* 20 (1975): 77-100; "Natural Selection and the Analysis of Human Sociality," in the *Changing Scenes in Natural Sciences,* 1776-1976. ed. C. E. Goulden, Philadelphia Academy of Natural Sciences Special Publication 12 (1977) 283-337. "Natural Selection and Social Exchange," in *Social Exchange in Developing Relationships,* ed. R. L. Burgess and T. L. Huston, (New York: Academic Press, 1978); "Evolution, Human Behavior, and Determinism," *Proceedings of the Biennial Meeting of the Philosophy of Science Association, 2 (1977): 3-21.*

5. Here I add a reservation about the enthusiasm which has occurred in the wake of Williams's 1966 book, and as a result of W. D. Hamilton's theory of inclusive fitness in "The Genetical Evolution of Social Behaviour, I, II," *Journal of Theoretical Biology,* 7 (1964): 1-52 (later referred to generally as "kin selection"). Hamilton followed Fisher *(The Genetical Theory of Natural Selection)* and others in pointing out that organisms can reproduce genetically not only via their direct descendants but also through whatever other nondescendant relatives may be socially available to them. In other words, nepotism to any genetic relative may be part of an organism's strategy of reproduction through altruism to others (altruism being defined as acts that at some expense or risk to the actor contribute to the well being—actually the reproduction—of others). Since humans are more extensively and complexly nepotistic than any other organism, Hamilton's arguments are immediately interesting to anyone concerned with human behavior. Hamilton's arguments also focused attention on subgenotypic elements, since they deal with reproductive costs and benefits to anyone of helping relatives with different fractional genetic overlaps. My caution has to do with the fact that evolutionary biologists have followed the lure of simplified quantitative genetics, perhaps without due care, right to the gene level. A prime illustration is the recent book by Richard Dawkins titled *The Selfish Gene* (New York: Oxford University Press, 1976). The same trend followed the rediscovery of Mendel's results, and that approach was eventually termed "bean bag genetics." As Ernst Mayr put it ("The Unity of the Genotype,"

Biologisches Zentralblatt, 94 (1975): 377-88), "The approach . . . became entirely atomistic and, for the sake of convenience, each gene was treated as if it were quite independent of all others. In due time all sorts of phenomena were discovered which contradicted this interpretation, such as the linkage of genes, epistasis, pleiotropy, and polygeny, and yet in evolutionary discussions only lip service was paid to these complications. . . . The purely analytical school thought that . . . an integrative attitude was incompatible with a meaningful analysis and dangerously close to such a stultifying concept as holism." Sooner or later, I believe, we must return to the individual as the most potent level at which selection works (See Lewontin, "The Units of Selection"). This simply means that, whatever we decide about subgenotypic levels, we will be forced to consider at every step what is meant by the fact that genes do not produce their effects independently of their genetic environments and are not inherited separately. This question may seem largely academic for social scientists, since the interests of genes and genotypes are so often synonymous; however, the prominence of certain questions, like kin selection and the outcome of parent-offspring conflict (see R. L. Trivers, "Parent-Offspring Conflict," *American Zoologist,* 14 [1974]: 249-64; Alexander, "The Evolution of Social Behavior") indicates that subgenomic considerations cannot be ignored, even by social scientists. Moreover, useful parallels can be drawn between the effects of selection on organization at subgenomic and supragenomic (i.e., social) levels (see Alexander and Borgia, "Group Selection, Altruism, and the Hierarchical Organization of Life."

6. The potential significance of this approach is amply illustrated by several recent studies showing that, while culture clearly is *potentially* independent of the interests of the genes, in fact cultural patterns in regard to social activities like birth-spacing, infanticide, reciprocity, war, inheritance, and interactions of genetic kin reflect to a surprising degree the genetic interests of their perpetrators (Alexander, "Natural Selection and the Analysis of Human Sociality"; chapters by various authors in Chagnon and Irons, *Evolutionary Biology.* . . .

7. Some of my students suggested two special cases that do not fit well into these general categories: (4) communal winter clusters (e.g., of flying squirrels) which may chiefly gain from minimizing energy loss, and (5) the V-formation of migrating waterfowl in which individuals may gain from pooling their information about the long migratory route. W. J. Freeland ("Pathogens and the Evolution of Primate Sociality," *Biotropica,* 8 (1976): 12-24) ar-

gues that disease may be an alternative cause of group living, but his arguments seem more to involve modifications of group living, once other causes establish and maintain it, because of the expense of diseases under group living.

8. Relative food shortages are here regarded as a "hostile force of nature," as Darwin also regarded them.

9. I use the term "function" in this paper essentially in the sense of Williams, *Adaptation and Natural Selection* (Princeton, N.J.: Princeton University Press, 1966) (and now of evolutionary biology more or less generally) to mean *evolved adaptive significance*. In other words, I use it to refer to any contributing factor supposed or hypothesized to be responsible for the selective origin and maintenance of the phenomenon, as opposed to (1) effects (Williams's "incidental effects") or (2) contributing causes unable by themselves to account for the phenomenon. Thus, there is an implicit assumption that what I am calling the *function* of an act or other phenotypic expression is alone capable of producing and maintaining the phenomenon; or that is an hypothesis under consideration. In this sense I am searching for single causes, and I regard this procedure as a logical approach to causation in biological phenomena. Multiple causes should be accepted only when single ones prove insufficient. If the absence of truly broad generalizations in the search for understanding of human existence should happen to be attributable to our long-term failure to reconcile the search with the principles of organic evolution, then our reluctance to admit the possibility of "single causes" should at least be tempered somewhat when we enter into a stage of rapid and massive incorporation of evolutionary principles, as I believe is the case at the moment.

10. Alexander, "The Search for an Evolutionary Philosophy of Man," pp. 115-17.

11. A. Keith. *A New Theory of Human Evolution.* (New York: Philosophical Library, 1949); Alexander "The Search for an Evolutionary Philosophy of Man"; R. D. Alexander and D. W. Tinkle, "A Comparative Book Review of *On Aggression* by Konrad Lorenz and *The Territorial Imperative* by Robert Ardrey," *Bioscience,* 18 (1968): 245-48; R. S. Bigelow, *The Dawn Warriors* (Boston: Little, Brown, 1969); R. L. Carneiro, "Slash- and Burn-Cultivation among the Kuikuru and Its Implications for Cultural Development in the Amazon Basin," in *The Evolution of Horticultural Systems in Native South America: Causes and Consequences; A Symposium,* ed. J. Wilbert, *Antropologica* (Venezuela), Suppl. 2 (1961): 47-67; E. O. Wilson, "On the Queerness of Social Evolution," *Bulletin of the Entomological Society of America,* 19 (1973): 20-22; E. O. Wilson,

Sociobiology: The New Synthesis. (Cambridge, Mass.: Harvard University Press, 1975); W. H. Durham, "Resource Competition and Human Aggression. Part I. A Review of Primitive War." *Quarterly Review of Biology*, 51, (1976): 385-415.

12. I am not implying that no other forces *influence* group sizes and structures but that balances of power provide the basic sizes and kinds of groups upon which secondary forces like resource distribution, population densities, agricultural and technological developments, and effects of diseases exert their influences.

13. For a discussion of the multiple consequences of this situation, see R. D. Alexander *et al*, in Chagnon and Irons, *Evolutionary Biology*. . . .

14. K. Flannery, "The Evolution of Civilizations." *Annual Review of Ecology and Systematics*, 8 (1972): 399-426. R. L. Carneiro, "Slash- and Burn-Cultivation . . ."; D. Webster, "Warfare and the Evolution of the State: a Reconsideration." *American Antiquity* 40 (1975): 464-70.

15. The criticism is sometimes made that Flannery's (ibid.) kind of reconstruction assumes that a particular modern ethnographic example is an exact replicate of its archaeological (and extinct) counterpart, or even, in the extreme, that the ethnographic and archaeological examples are implied to give rise to one another, always progressing from simple to complex. This attitude implies a basic misunderstanding of comparative method. Comparative method, in biology or archaeology, assumes: (1) that sequences of change have occurred (genetic evolution and cultural change); (2) that parallel sequences of change occur in different places, at different times, and in different lines at the same times and places; (3) that some (but not all) of the attributes of different stages (but not the actual cases or even, necessarily, the actual sequences) will be represented in both extant and extinct forms, and (4) that appropriate comparisons of such attributes can yield information about the sequences of change and their causes. These assumptions allow interpretation of the past by studying the present, or vice versa, and comparative method, explicitly in the sense described here, represents the main source of evidence for both evolutionary biology and archaeology.

16. There *are* genetic and physiological constraints on natural selection: They are recognized by evolutionists under the term "specialization." An animal like a mole, specialized to live underground, is less likely to evolve wings than one, like a squirrel, which spends its time climbing and leaping from tree to tree. This is a very simple example, but the argument is essentially the same whether

one is considering subgenomic interactions or populational phenomena.

17. H. W. Power, "On Forces of Selection in the Evolution of Mating Types. *American Naturalist,* 110 (1976): 937-44.
18. Carneiro, "Slash- and Burn-Cultivation . . .," actually approached this argument with his concepts of environmental and social circumscription.
19. See Fisher, *The Genetical Theory . . .* on "Heroism and the Higher Human Faculties."
20. H. Kelsen. *What is Justice? Justice, Law, and Politics in the Mirror of Science; Collected Essays.* (Berkeley: University of California Press, 1957).
21. I am in no way arguing that all humans always behave so as to maximize reproduction but I am arguing that this is what they have *evolved* to do, in the *environments of* history, and that we must know ourselves in this way to understand best all of our inclinations and our motivations.
22. I would expect the function of laws (see footnote 9) to be the limitation of the reproductive striving of those *other than* the legislators and enforcers themselves; it is an incidental effect that legislators and enforcers are limited by the same laws—although from the viewpoint of those requiring legislators and enforcers to follow the same laws they have to follow (another form of enforcement), this "effect" in turn becomes a function.
23. R. D. Alexander *et al.,* "Sexual Dimorphisms and Breeding Systems . . ."
24. E. H. Sutherland and R. Cressey, *Principles of Criminology,* 7th ed. (New York: Lippincott, 1966), p. 26, note that: "The crime rate for men is greatly in excess of the rate for women—in all nations, all communities within a nation, all age groups, all periods of history for which organized statistics are available, and for all types of crime except those peculiar to women, such as infanticide and abortion."
25. Moreover, only very recently (e.g., Michigan Supreme Court ruling, 1977) has it been suggested formally that a woman has the right to choose sexual partners, in the sense that her sexual behavior in general cannot be used against her in court proceedings testing whether or not, in a specific instance, she has been raped.
26. D. J. Mulvihill and M. M. Tumin, *Crimes of Violence* (Washington, D.C.: U.S. Government Printing Office, 1969).
27. N. A. Chagnon, *Yanomamö: The Fierce People* (New York: Holt, Rinehart, and Winston, 1968); J. Himelhoch, "A Psychosocial Model for the Reduction of Lower-Class Youth Crime," in R. L.

Akers and E. Sagarin, eds., *Crime Prevention and Social Control* (New York: Praeger, 1972) pp. 3-14; Sutherland and Cressey, *Principles of Criminology*; Mulvihill and Tumin, *Crimes of Violence*.

28. F. Ferracuti and S. Dinitz. "Cross-cultural Aspects of Delinquent and Criminal Behavior," in M. Reidel and T. P. Thornberry, eds, *Crime and Delinquency: Dimensions of Deviance*. (New York: Praeger, 1974), pp. 18-34; see also B. M. Fleisher, *The Economics of Delinquency* (New York: Quadrangle Books 1966).

29. F. Ferracuti and S. Dinitz, "Cross-cultural Aspects . . ."; J. Himelhoch, "A Psychosocial model . . ."; J. B. Cortes and F. M. Gatti, *Delinquency and Crime: A Bio-psychosocial Approach*. (New York: Seminar Press, 1972); J. P. Clark and E. P. Wenninger, "Socioeconomic Class and Area as Correlates of Illegal Behavior among Juveniles." *American Sociological Review*, 27 (1962): 826-34. See also B. M. Fleisher, *The Economics of Delinquency*.

30. R. A. Cloward and L. E. Ohlin, *Delinquency and Opportunity* (Glencoe, Ill.: The Free Press, 1960); Fleisher, *Economics of Delinquency*; T. Hirschi, *Causes of Delinquency* (Berkeley: University of California Press, 1969); L. Radsinowski and M. E. Wolfgang, eds., *Crime and Justice. The Criminal in Society*, vol. 1; (New York: Basic Books, 1971); C. A. Hartjen, *Crime and Criminalization (New York: Praeger, 1974)*.

31. Hamilton, "The Genetical Evolution . . ."; R. L. Trivers, "The Evolution of Reciprocal Altruism," 1971; M. J. West Eberhard, "The Evolution of Social Behavior by Kin Selection," *Quarterly Review of Biology*, 50 (1975): 1-33; Alexander, "The Evolution of Social Behavior"; "Natural Selection and the Analysis of Human Sociality" 1977; "Natural Selection and Social Exchange" 1978.

32. Trivers, ibid.; Alexander, ibid.

33. Alexander, ibid.; Alexander and Borgia, "Natural Selection and the Hierarchical Organization of Life."

34. Flannery, "The Evolution of Civilizations."

35. P. Wiessner, *Hxaro: A Regional System of Reciprocity among the !Kung San for Reducing Risk*. Ph.D. Thesis, University of Michigan (1977).

36. Flannery, "The Evolution of Civilizations."

37. N. A. Chagon, *Yanomamö: The Fierce People;* "Genealogy, Solidarity, and Relatedness: Limits to Local Group Size and Patterns of Fission in Expanding Populations." *Yearbook of Physical Anthropology*, 19 (1975): 95-100; (1974).

38. Flannery, "The Evolution of Civilizations."

39. Flannery, Ibid.

40. P. Stein and J. Shand, *Legal Values in Western Society*. (Edinburgh: Edinburgh University Press, 1974), pp. 114-16, provide a closely parallel but evidently independent comment: "In the fellowship type of social relationship, the value of the individual as a person is secured by the mutual regard and affection of the members for each other. The nature of the relationship is such that every member can confidently rely on receiving respect from the others. Their mutual regard is the product of its personal character. Such a social group cushions its members against the impact of legal rules. For example, early Celtic society, which was largely pastoral, displayed marked fellowship features. The main social unit was the kindred, the *drebfhine* of Ireland and Gaelic Scotland, which extended for four generations. The act of one individual might affect all the members of the kindred, each could claim his share in any inheritance, and each was bound to assume his share of liability for any fines payable by any member. As Nora Chadwick says:

> There was no personal payment. The 'kindred' stood or fell together. In this way they were responsible for one another and would obviously keep a close eye on one another's doings. In this way too every 'kindred' group would see to it that the kindred did duty as both police and judges. There could have been no better way in such a society of keeping justice on an even keel, and this helps to explain the relative scarcity of legal machinery which a study of so many legal tracts implies.

"Traces of such group feeling can be found even today in closely knit family groups, which regard themselves as culturally distinct from the rest of society. Gypsies, for example, settle their disputes themselves according to their own customs, and will rarely have recourse to law, except in their dealings with outsiders. If a member of a gypsy family in East Anglia is accused of a motoring offence, it is common for his whole family to accompany him to court, and if he is convicted and fined they will all contribute as a matter of course to its payment.

"As societies develop into the nation-states, they cease to be collections of fellowship groups. These groups are replaced by less personal types of social relationship, in which the members feel no special regard for each other. In the newer relationships respect for persons cannot be taken for granted. Circumstances require that people be treated as individuals, and the position of the individual in society must be recognised by the law. Further, the precise character of the law is best adapted to a society whose members are treated as separate individuals rather than as members of groups. Historically, as laws have become more sophisticated, the more they have tended to make the individual rather than the group the focus of rights and duties. These considerations do not, however, imply the attribution of a particular value to the individual as against society.

"Ancient Roman society regarded property as belonging to the family, but quite early in its development it ascribed ownership to the head of the family, the *paterfamilias*. He could dispose of the family farm, for example, without the need to obtain the consent of other members of the family. The freedom of disposition applied both to alienations *inter vivos,* such as followed a sale of the property, and to those by will. Towards the end of the Republic, it is true, a testator was compelled to take into account the needs of his descendants when deciding the destination of his property after his death, but he was still allowed a very wide discretion. This aspect of the Roman law of property is sometimes cited as evidence of Roman dedication to the principle of individualism in the modern sense. Such an assumption is unwarranted. The freedom of disposition enjoyed by the Roman *paterfamilias* was legally and commercially convenient. Its exercise must be seen against the background of the strong social pressures of good faith, family piety, and neighbourly duty, summed up in the notion of *officium.* These pressures considerably inhibited the use which the *paterfamilias* made of his powers of disposition. Furthermore, what the Roman owner could do with his property, apart from his rights of disposition, was not so unrestricted as it has in modern times been declared to be. As we shall see, Roman law kept Roman owners within the limits of good neighbourliness, and the alleged 'absolute' character of Roman *dominium* has largely been read into the Roman texts by later generations of jurists imbued with non-Roman ideas. Had the Romans really been individualistic in the modern sense, they would have changed the rules whereby adult descendants, unless formally emancipated from the power of the *paterfamilias,* could own no property of their own in his lifetime. The law recognised their right to control what had come to them as a result of their own enterprise, such as military service, but anything they received by way of legacy or gift belonged to the *paterfamilias,* and was thus kept in the family funds of which he disposed."

41. I mean that our view of evolution may parallel our view of pornography, through reflections, conscious or not-so-conscious, about the effects of either on our children. Thus, we seem first to teach our children to be absolutely truthful—a way of operating that

clearly is incompatible with social, economic, and political success, probably in any society anywhere. *Then* we teach them to adjust the truth ever so slightly—and thereby successfully—to their own advantage. After they start telling Aunt Kate that she is fat, and such things, we begin to teach them to be what we so tactfully call tactful. Similarly, we seem first to teach our children that sex is an evil to be avoided—that too is a way of operating, for adults, at least, that is not usually compatible with either social or reproductive success. *Then* we teach them, or allow to develop, the circumstances in which sex is permissible and profitable; we allow them to learn that sex in these situations is enjoyable, and we try to teach them that sex in other situations is not. Perhaps, in each case, the sequence of learning is crucial; perhaps by these sequences alone we are able reliably to guide children to success in the sensitive business of sociality, sexuality, and morality. In each case a connection can be discerned between the education of children and the growth of understanding about evolution. In each case a possible explanation of the relationship between family stability and law breaking is discernible. Perhaps, without always being conscious of it, we tend to be repelled by evolutionary explanations, particularly of human sociality, because we somehow understand that full knowledge and acceptance of them would not be good for our children (see also R. D. Alexander. "Creation, Evolution, and Biology Teaching," *American Biology Teacher,* (Feb. 1978).

42. Donald Black's book, *The Behavior of Law* (New York: Academic Press, 1976) did not become available to me until the final draft of this manuscript was completed. Black's findings are relevant to my arguments and seem to support (even if inadvertently) the general viewpoint I have advocated. As his title suggests, for purposes of analysis, Black treats law as Leslie White, e.g., *The Science of Culture,* (New York: Farrar, Straus, 1949) treated culture—as a thing apart from function, motivations, psychology, and individuals. He seeks correlates of the *quantity* of law (pp. 6-8) and tries to ascertain their effects. He defines law as "governmental social control" (p. 2), and quantifies it chiefly (p. 3) by "the number and scope of prohibitions, obligations, and other standards to which people are subject, and by the rate of legislation, litigation, and adjudication." He then examines the correlates of quantitative variations in law in different circumstances and societies, and emphasizes twenty-five or thirty such correlates, which may be condensed as follows: Law is "greater" (employed more often, or more effectively) in societies and social groups that are larger, more dense, more organized, more differentiated, more complex, more

stratified, and in circumstances in which there are fewer other social controls (e.g., less family control) and greater "relational" (social, genetic) distances among interactants (e.g., more during interactions between distant relatives, or nonrelatives, and "strangers"), than in the opposite kinds of societies, social groups, and circumstances. Within societies "more" law is directed (or law is directed more often and more effectively) at individuals and groups that are relatively low-ranking, uninfluential, transient, not "respectable," socially marginal, and more distantly related then in the opposite direction.

Black's approach treats law as a singular phenomenon whose traits can be analyzed and generalized. Because law is obviously not without function, and is not independent of the motivations of people, Black's success in locating a small number of general rules, despite the enormous variation in legal systems, suggests that a certain singularity of function, therefore of motivational background, may exist for law as a whole. That is also the argument made in my paper. Moreover, the particular correlates discovered by Black sometimes are the same as those I have emphasized, and his findings seem to support the arguments about the origins and functions of law described in my paper.

Commentary

Sociobiology and Evolving Legal Systems: Response to Richard D. Alexander

Kenneth F. Schaffner

I. Introduction

Professor Alexander's essay is an inquiry into the light that modern construals of Darwinian evolutionary theory can shed on the development and nature of societal laws. Alexander notes that he wishes "to utilize the approach of generalizing and predicting from the process of adaptation as a vehicle for studying human sociality, especially to consider the nature and probable background of societal laws, norms, and traditions" (p. 250).[1] Alexander believes that we *must* pursue this type of inquiry, and he writes "whether we like it or not, . . . we are required to accept that *background explanations for all activities of life including our own behavior, will eventually be found in generalizations deriving from cumulative effects of an inevitable and continuing process of differential reproduction of variants*" (pp. 253-54).

Later on in his essay, after considering various prime movers of human social evolution—a matter to which I shall return in the next section—Alexander makes several most interesting and dramatic claims. "The kind of argument I am making here," he writes, "cannot fail to be disconcerting, or even bizarre, to many modern scientists, philosophers, and humanists, especially those who are satisfied with their present way of looking at things; it is too novel for anything else to be the case" (p. 261). He adds that he believes that "application of Williams's refinement of Darwinism, as I am attempting to do here, seriously threatens current philosophical thinking at its base and cannot fail to alter dramatically the theoretical underpinning of the social sciences . . . in the sense that interpretations of history, and predictions about the future of humans not yet cognizant of these matters, will be facilitated by this kind of thinking more than by any other aspect of human understanding" (pp. 261-62).

Let us look at the arguments that support this almost imperialistic claim for evolutionary biology.

II. The Outlines of Alexander's Thesis

There is, in my view, a curious dualistic aspect to Professor Alexander's view about the evolution of societal laws. The two components of his evolutionary model for societal laws are not necessarily incompatible, but they are not well harmonized in his presentation. (The two components may even be incoherent, but it is difficult to determine this because the components are characterized in rather vague terminology.)

One component of Alexander's model is the war-balance of power, "prime mover" of human evolution. In the early part of his paper, Alexander reviews all the "benefits" which might accrue to individual organisms as a result of group living. He discerns three such main benefits:

1. predator protection either because of (a) group defense or (b) the opportunity to cause some other individuals to be more available to the predator;
2. nutritional gains when utilizing food such as (a) large game, difficult to capture individually, or (b) clumped food difficult to locate; and,
3. simple crowding on clumped resources. (p. 255)

Benefits are construed as benefits to the individual in line with Alexander's acceptance of George C. Williams's thesis that selection is at the genic or individual level,[2] but not at the group level, a point to which I shall return in later sections. Alexander rejects alternatives (2) and (3) and elaborates the concept of predator protection in human groups to include both protection against other human groups and also a balance of power between such groups. What I term the first component of Alexander's model, then, is the premise that a combination of war and balances of power is the prime mover of human evolution.

The second component of Alexander's model is developed in his discussion of societal values and of justice. Here Alexander presents a eudaimonistic descriptive ethics, proposing that what individuals proximately strive for is happiness and pleasure. Evolutionary biological insights, however, tell us that happiness is "an evolved means to an end." Alexander writes that "pleasure and happiness associate with events and stimuli that are beneficial to us in the usual environments of history" (p. 264), and " 'beneficial' is defined, in terms of history, as leading to reproduction, i.e., as leading to genetic survival" (p. 264). Our ultimate interest then, according to Alexander, is in reproduction, and it is in terms of this second component—*reproductive striving*—that Alexander analyzes the evolution of societal laws. Alexander writes that:

> The basis for conflicts of interest among individuals—hence the basis for the unresolvability of the question of justice—evidently derives from our history of reproductive competition acting primarily at the individual level. I am saying bluntly that social conflict derives from biological facts. (p. 266)

Alexander then adds another premise—that the main function of societal law is to preserve order and, in the light of the second component of his model, this translates into his interesting claim that "*the function of laws is to regulate and render finite the reproductive strivings of individuals and subgroups within societies, in the interests of preserving unity in the larger group* (all of "society" or the nation-state). Presumably, unity in the larger group feeds back beneficial effects on those segments or units which propose, maintain, adjust, and enforce the laws." (p. 267).

Alexander then goes on to test this hypothesis by looking for correlations between groups and stages of development representing peaks of reproductive striving, and law-breaking behavior.

A very brief section on "Nepotism and Reciprocity" follows in which it is asserted but not argued that "nepotistic behavior toward nondescendant relatives evolves out of parent-offspring interactions, and that reciprocity derives from nepotism" (p. 272). (This derivation is presumably important since it would tie the two general classes of social interaction to an evolutionary biological base of maximizing reproductive self-interest, but I have serious doubts that it can be made.)

The following section traces "Changes in Rules with Development of Nation-States." Alexander acknowledges that this section is indebted to Flannery's account of the rise of the nation-state through the stages from bands to tribes to chiefdoms to states.[3] Though I will in general restrict my comments to the next sections, I cannot pass without noting that there is very little in this account laying out detailed examples of rules and relating specific changes in the rules to the twin movers of war-balance of power and of reproductive striving.

Alexander does present a tripartite classification of the laws of nation-states which he believes is explained by the reproductive striving hypothesis. Laws against murders, assault, rape, kidnapping, treason, theft, extortion, and breach of contract are viewed as restrictions which "prevent individuals or groups from too severely interfering with the reproductive success of others" (p. 276). Laws against polygamy, nepotism, tax evasion, draft evasion, and monopolies are construed as restrictions preventing "individuals or groups from too dramatically enhancing their own reproductive success" (p. 276). Finally, laws concerning patents, copyrights, and wills are concerned with promoting "industry and creativity in individuals and groups in ways that may be exploited or plagiarized by the larger collection" (p. 276).

Professor Alexander ends his paper with some comments on his view of the relation between evolutionary biology and normative ethics to which I shall return later after critically analyzing the main argument just outlined.

III. Criticisms of the Main Argument and Its Assumptions

At the beginning of the previous section I noted that what I discerned as the two main components of his model of societal evolution were both rather vaguely defined and were not well harmonized, in the sense that the two components were not shown working together in a joint explanatory process. To be more specific on the first point, it would appear that the striving-for-reproduction is an active force which is constrained and modified by the other prime mover, the war-balance of power component. Both of these "movers" are extremely general terms, and each presumably admits of multiple forms of realization and interaction. In order to develop a set of concepts that are clearly coherent and sufficiently precise to be incorporated in testable hypotheses, more exact specification of what is meant by reproductive striving and balance of power is required. Otherwise the terms are, I believe, sufficiently vague and elastic to account for almost *any* conceivable historical development of societal laws. The point I want to make here then is that more precision and detail is needed in conceptually clarifying the foundation of Professor Alexander's theory.

The second point I want to consider is the coherence of the two components of (1) war-balance of power, and (2) reproductive striving. In his section on "Group Living and Rules" (pp. 262-63), Professor Alexander suggests that rules (or laws) are introduced because they benefit the individual and the group. After reviewing certain complexities associated with the notion of justice, Alexander concludes that the benefits introduced by laws are a consequence of the law's main function to preserve order, or more specifically, *"to regulate and render finite the reproductive strivings of individuals and subgroups within societies, in the interests of preserving unity in the larger group . . ."* (p. 267).

Now presumably this crucially important, for Alexander, function of law could be related to the prime mover war-balance of power, which would act as a cause or shaper of the law. This, however, is not worked out in any detail in Alexander's essay. In addition, there is no evidence provided that the regulation of the reproductive strivings of individuals and subgroups is increased in circumstances of war or external threats. In point of fact, our intuitions lead us to suspect the reverse. What we seem to have then in Alexander's model are two prime movers which *may* function coherently to account for the rise and modification of law, but which are not shown to do so.

A third point of criticism I want to raise is in connection with two methodological and one substantive premises of Alexander's argument. As noted in Section II above, Alexander has accepted wholeheartedly George C. Williams's thesis that one ought to construct evolutionary arguments in the simplest terms or on the lowest level possible—at the level of the gene or at most the level of the individual. Alexander accepts this methodological assumption as one of his premises—he does not argue for it but refers us to Williams's (and also Lewontin's) work. (In point of fact, however, Lewontin's data only supports selection at the *level of the chromosome*.)[4]

Now to me, this begs the question in an important way, and I think it also introduces incoherence into Alexander's own view of the evolution of societal laws. I shall come back to this latter point below, but suffice it for now to note that other evolutionary biologists, such as E.O. Wilson, think this methodological assumption of Williams's is too restrictive and not yet proven. Wilson points out in his *Sociobiology* that "group selection and higher levels of organization, however intuitively improbable they may seem, are at least theoretically possible under a wide range of conditions. The goal of investigation (in sociobiology) should not be to advocate the simplest explanation, but rather to enumerate all of the possible explanations improbable as well as likely, and then to devise tests to eliminate some of them."[5]

There is a second methodological assumption in Alexander's approach which I think also needs additional defense. This is his preference for a "singular" or unifactorial explanation of human behavior (p. 260). The only argument given here relies on the (questionable) acceptance of Williams's thesis noted above. Accordingly it would seem that a more sympathetic consideration of multifactorial explanations of social evolution might be warranted.[6]

This point brings me to my comment on the substantive assumption which Alexander makes regarding a combination of war and balance of power as the prime mover of human social evolution. Alexander's summary of possible agents of evolution was partially restricted by his acceptance of Williams's methodological assumption, and further supported by Alexander's preference for a singular explanation. It should be noted, however, that additional factors to those which Alexander considers have been invoked in the sociobiological literature to account for social evolution. Professor Alexander does not discuss the possibility of sexual selection, of multiplier effects on cultural innovation and in network expansion, or the effect of the development of agriculture. Other writers, for example Adams,[7] have also preferred more complex multifactorial accounts of the state.

What I am urging here is a more explicit comparative mode of argument than Alexander gives. One *could* rejoin that systematic development of one plausible thesis about the evolution of societal laws is sufficient, but it seems to me not to be the case. For in an area where one is at least initially (1) dealing with vague concepts, (2) proceeding with little control by a general theory—note the lack of *any* population genetics arguments in Alexander's essay, and (3) lacking broad paleontological evidence to serve as an empirical control over speculation, it is methodologically desirable that one should proceed *comparatively,* considering the strengths of alternative theories in offering "as good" or better explanations. It would seem that only in this way can we ferret out the weaker speculative claims and put more effort into conceptually clarifying and seeking additional evidence to test the more plausible approaches.

Let me now return to a point I mentioned briefly above. I suggested that Alexander's view of law might not be coherent with his acceptance of Williams's assumption rejecting group selection. As I noted above on p. 295, Alexander's view of the role of law is "to regulate and render finite the reproductive strivings of individuals." Now if Alexander is willing to admit that the individual is the unit of selection, we must remember that there are *intra*-individual control systems that regulate and render finite the reproductive strivings of an individual's *cells.* When a cell escapes the normal constraints of genetic control (and immunological surveillance), and proliferates in an uncontrolled way, the result is a cancer. The parallel should be obvious. If law regulates individuals in a society, modifying their behavior in significant ways, I do not see that there is not sufficient cohesion of a law-regulated social group that the group can count as an individual in Williams's sense.[8] To suppose otherwise is to subscribe to a kind of "beanbag sociobiology," to use a modification of Ernst Mayr's felicitous phrase,[9] in which only the actions of individual organisms are considered.

This view would, I think, introduce the law as a new, partially emergent factor affecting social evolution. This is not a novel thesis and it has been held by a number of authors, though not by Alexander. It is sufficiently interesting and yet also philosophically troublesome, because of the complexities of the notion of "emergence."

IV. Emergent Social Evolution?

It has often been suggested that human evolution over the last hundred thousand years or so is strongly conditioned by cultural factors which are importantly decoupled from their biological base. The late eminent geneticist, Dobzhansky, for example, wrote in 1963 that:

> Culture is not inherited through genes—in a sense human genes have surrendered their primacy in human evolution to an entirely new non-biological or superorganic agent, culture. However, it should not be forgotten that this agent is entirely dependent on the human genotype.[10]

Now I would like to argue by analogy with the relation between chemistry and biology, that we should give serious consideration to the view that cultural evolution involves a set of interactions *originally arising out of a biologically determined ground,* but which has for a long period been severely *underdetermined* by *purely* biological principles. The biological underdetermination is sufficiently extensive, I would speculate, that we will obtain at best very partial explanations of social evolution by sociobiology in its purely biological aspects.

Now I advance this qualified emergentist thesis as a speculative one—as one alternative to a biological determinism. To make it plausible let me turn to a discussion of the parallel with DNA, genes, and chromosomes.

The "qualified emergentism" which I should like to urge is based on an analogy with the relation of biology to chemistry. In

1967 Michael Polanyi[11] and I[12] independently advanced the thesis that "boundary conditions" (Polanyi) or "initial conditions" (Schaffner) in extant biological systems caused problems for reduction. Briefly put, these initial conditions describe the *organization* of the chemical components—and this organization is not dictated by purely chemical constraints except as those constraints are provided by templates given by already living systems. An example from molecular genetics may help in making the point clear. From a *purely chemical* perspective, there are no constraints on the sequence of DNA nucleotides, namely of adenine (A), thymine (T), cytosine (C) and guanine (G) on a single strand of a DNA helix. The sequence CAATG . . . is as chemically stable as GCGAT . . ., or any other such sequence. This was noted by Watson and Crick in their original paper on the double helix structure for DNA, where they wrote, "The sequence of bases on a single chain does not appear to be restricted in any way."[13] Specific sequences of DNA daughter molecules are, however, dictated by the (complementary) sequence of the parent strands: the parent strands serve as templates for the synthesis of specific sequences of nucleotides which in strings of several hundred nucleotides constitute genes. (The genes are further combined into chromosomes.) The organization that constitutes the organism's genome then "supervenes" on the simple chemical rules governing DNA sequences, even through the organization is statable in chemical terms.

Polanyi and I drew radically different implications from the existence of this organization, he seeing a series of levels of irreducible organizing principles which harnessed the underlying material. I, on the other hand, saw the need for a *chemical* evolutionary theory to explain reductionistically the historical genesis of these supra-chemical constraints.[14] Subsequent discussion with Polanyi did not lead to a resolution of these divergent interpretations: he seemed to distrust Darwinian evolution in general and chemical evolution in particular.

I would now like to suggest that something analogous to the relation between biology and chemistry is at work in the relation between sociocultural disciplines and the biological sciences. The conclusion of the argument is that we must take sociocultural organizing principles as at present given, and that these, from a *current* point of view, "supervene" on the basic biological systems including behavioral dispositions. One crucially important set of such sociocultural organizing principles is the set of *laws* that regulate societies. The stages of my argument are that (1) evolutionary explanations in biological terms i.e., in genetic, phenotypic, and selection pressure terms, are only partially reconstructable. Fossil records are fragmentary and behavior traits do not leave strong paleontological traces. (2) Social organizations of organisms are likely due to a combination of genetic determinism and stochastic behavioral innovations which can become fixed by mimicry, and these innovations can in turn exercise an important influence on the evolutionary adaptability of populations. These innovations thus can result in sociocultural modifications which, because of both their stochastic origin and the gappiness of the evolutionary record, are best treated or characterized in sociocultural terms, conditioned, but not fully determined in purely biological terms by the underlying biological systems. (3) Sociocultural evolution should accordingly be treated as a set of forces partially decoupled from the biological base. Explanations of a group's behavior may well involve both sociocultural factors and biological factors, but from what is possible on the basis of currently available information (and I think all information available in the future), one must utilize this dual, nonreductionist approach in accounting for social evolution. This view bears some resemblance to what Edward Wilson has termed, following earlier ideas of Pringle, Bateson, Skinner, Levins, and others, a partially decoupled hierarchial tracking system.[15]

This dual nonreductionistic approach is functionally equivalent to a qualified emergentism. The emergentism is qualified for two reasons. First, *if* the *complete* evolutionary record were available, then I would be disposed toward the likelihood that the behavior of humans in society could be explained as a consequence of evolution operating over millions (if not billions) of years. Second, the emergentism is qualified because even in the absence of the complete evolutionary record, the view outlined here does not argue against the *in principle* reducibility of human behavior (including sociocultural behavior) to physics and chemistry, when the extraordinarily complex physicochemical initial conditions are added to the reducing science's general laws, models, and theories.

V. Evolutionary Biology and Normative Ethics

In this last section of my comments, I would like to respond to some of the provocative ideas which Professor Alexander advances concerning the relation between his view of evolutionary biology and normative ethics.

It strikes me that Alexander is taking both a too pessimistic and a too optimistic view of the implications of an evolutionary theory of ethics. He suggests that humans are "sufficiently plastic in their behavior to accomplish almost *whatever they wish*" (pp. 277-78). This is too optimistic—the thesis is not defended and anyone but the most complacent conservative would, I think, dispute it. Genetic and social inertia are powerful constraints on the plasticity of behavior.

This is also why I think Alexander is too pessimistic in his view of the irrelevance of evolutionary biology for normative ethics. Though I have severe doubts about the current state of the intellectual credentials of a sociobiology extended to human behavior, in principle sociobiology could tell us important facts about human disposition and potentialities. A theory in normative ethics is rarely if ever fully insulated from factual implications. The consequences of eliminating "pleasure" from utilitarian theories,[16] for example, should suggest that there are interactions between ethical theories and at least a philosophical anthropology with partial factual pretensions.

Evolutionary biology might also help normative ethics because it may provide deeper and more far-ranging explanations of human behavior than explanation in terms of motivation by "proximate rewards." Alexander thinks that "proximate rewards" are the basis of normative ethics. A deontologist such as Kant would question this,[17] but I believe that even a deontologist who was not commited to an a priori ethical theory would find a potentially deeper knowledge of the behavior of humans and groups of humans helpfully relevant to formulating an ethical theory.

NOTES

1. Page references given in text are to Professor Alexander's essay in this volume.
2. G.C. Williams, *Adaptation and Natural Selection* (Princeton: Princeton University Press, 1966).
3. K.V. Flannery, "The Cultural Evolution of Civilizations," *Annual Review of Ecology and Systematics,* 3 (1972): 399-426.
4. Alexander cites G.C. Williams, *Adaptation and Natural Selection,* and R.C. Lewontin, "The Units of Selection," *Annual Review of Ecology and Systematics* 1 (1970): 1-18. But see I. Franklin and R.C. Lewontin, "Is the Gene the Unit of Selection?" Genetics, 65: 707-35, and R.C. Lewontin *The Genetic Basis of Evolutionary Change* (New York: Columbia University Press, 1974), esp. chap. 6, for data and arguments implicating the whole chromosome as the unit of selection.
5. E.O. Wilson, *Sociobiology: The New Synthesis* (Cambridge: Harvard University Press, 1975), p. 30.
6. In fairness to Alexander, it should be pointed out that in footnote twelve of his essay he admits that other factors in addition to the war-balance of power "prime mover" may affect social evolution. He views these factors as secondary, however, which exert some rather small influences on the main balance of power factor.
7. R. McC. Adams, *The Evolution of Urban Society* (Chicago: Aldine Publishing Co., 1966).
8. For a discussion of a related notion of species as individuals see M.T. Ghiselin, "A Radical Solution to the Species Problem," *Systematic Zoology* 23 (1975): 536-44 and L. Van Valen, "Individualistic Classes," *Philosophy of Science,* 43 (1976): 539-41.
9. See E. Mayr, "The Unity of the Genotype," *Biologisches Zentralblatt,* 94 (1975): 377-88, as quoted by Alexander, in footnote five of his essay.
10. T. Dobzhansky, "Anthropology and the Natural Sciences—The Problem of Human Evolution," *Current Anthropology* 4 (1963): 146-48.
11. M. Polanyi, "Life Transcending Physics and Chemistry," *Chemical Engineering News* 45 (1967): 54-7.
12. K.F. Schaffner, "Antireductionsim and Molecular Biology," *Science* 157 (1967): 644-47.
13. J.S. Watson and F.H.C. Crick, "A Structure for Deoxyribose Nucleic Acid," *Nature* 17 (1953): 737.

14. K.F. Schaffner, "Chemical Systems and Chemical Evolution," *Amerian Scientist* 57 (1969): 410-20.
15. Cf. Wilson, *Sociobiology,* chaps. 7 and 27.
16. Cf. J.S. Mill, *Utilitarianism* (London: Longmans, 1907).
17. Cf. I. Kant, *Foundations of the Metaphysic of Morals:* Text and Critical Essays. ed. Robert P. Wolff. (New York: Bobbs-Merrill, 1969) [originally published in 1785].

Commentary

Rejoinder to Kenneth Schaffner

Richard D. Alexander

SCHAFFNER DESCRIBES my opening remarks about the significance of evolution for human understanding as "an almost imperialistic claim for evolutionary biology." But knowledge advances through a competition of ideas not disciplines. It is not an imperialistic claim of physics that humans are subject to gravity. To doubt my arguments is appropriate; to imply that their acceptance would constitute a supplication to the discipline from which I operate is not.

Schaffner is disturbed by the "absence of argument" in my paper about the relationships between (1) parent-offspring interactions and other forms of nepotism, and (2) altruism to relatives and reciprocal altruism between nonrelatives. Actually, my footnote 31 cites several references in which these relationships are discussed. Here I especially refer the reader to my paper on social exchange where I argue that equity and exchange theory in sociology and psychology have encountered difficulties with what investigators in these fields have termed "deep" and "intimate" interactions because they have failed to recognize that these are either nepotistic or (because of novel social situations) surrogate-nepotistic in nature.[3] Thus, like the parent-offspring interaction, they have in the past involved genetic returns that nullified the requirement of *phenotypic* equity in exchange. For most organisms, parenthood is wholly an expense to the parental phenotype;

so is nepotism in general, with the payoff in both cases genetic. In reciprocal or exchange relationships the payoff is phenotypic. Nevertheless, in evolutionary terms the altruism in all of these interactions is similar in that (1) it represents an investment with risk, and (2) in parenthood or nepotism it may involve both genetic returns and phenotypic reciprocity.

Schaffner probably means to suggest that if social reciprocity can be shown to have no historical or other connections to nepotism, then perhaps the human social interactions that appear to be cemented by reciprocity (as I have suggested for the unity of large nations) can be inferred to be independent of biological history. All that is required to deny this suggestion is some degree of intrication of nepotism and reciprocity in human affairs. Scarcely anything shows better than does the structure of law that such intrication is the rule in everyday life. Moreover, social reciprocity can arise on strictly selective grounds in the absence of nepotism;[2] after all, one of the oldest social interactions (♂-♀) begins as reciprocity. Schaffner's objection parallels the incomplete view of social reciprocity expressed by Sahlins,[3] in which he asserts that nonrelatives cannot gain biologically from social interactions; I have explained Sahlins' error elsewhere.[4]

Schaffner attempts to weaken my argument about the development of nation-states by describing it as "heavily indebted" to a single account—that of Flannery.[5] But Flannery's paper is a review article which the reader can use to direct himself to the anthropological literature on this subject. Flannery does not suggest that his reconstruction is unique; instead it is solidly backed by the work of numerous anthropologists and archaeologists.

Contrary to Schaffner's implication, the concept of "reproductive effort" is firmly established in biology and is essential to the explanation of the finiteness and the patterning of individual lifetimes.[6] For a review of the concept and its theoretical basis I suggest Hirshfield and Tinkle;[7] for a discussion of the kinds of effort shown by organisms, their variations between male and female and between species with different kinds of breeding systems, see Alexander and Borgia.[8]

Schaffner says that my use of the idea of "balances of power" is vague. But I see this notion as pervading every imaginable aspect of human existence, with its role understandable by ordinary humans in everyday circumstances that involve individuals as well as groups of all sizes, compositions, and purposes. I cannot imagine anyone who understands the evening news or who functions satisfactorily in ordinary social interactions viewing the concept of balances of power as vague.

In connection with the problem of levels in the organization of life at which natural selection is most effective, Schaffner seems to have interpreted the rule of parsimony to mean that we must always invoke selection at the lowest *possible* level. This is a misinterpretation: On parsimony we invoke selection at the lowest *acceptable* level, that is, the lowest level at which, on the basis of available information, it seems to explain things. The reason for using the *lowest* (in this sense) level is simply that fewer assumptions are involved.

Schaffner castigates me for preferring singular explanations.[9] But all valid principles and generalizations amount to "singular explanations." Accordingly, I seek them unabashedly, and when they work I prefer them.

Schaffner accepts Wilson's argument against "advocacy" of one among several possible explanations. But Wilson's argument is inconsistent because it is itself an advocacy of a particular hypothesis about procedures. More than this, it is advocacy of a silly and wasteful procedure. He indicates that merely because one "explanation" is more probably correct (one can only presume that this means "superior on the available evidence") is no excuse for "advocating" (using?) it, and that we must instead, each time, review all possible explanations and eliminate them. But hypotheses cannot really be eliminated and conclusions are never absolute. All one can do with hypotheses or conclusions is to continue to the next step in analysis.

The problems of "levels" and "units" of selection is an extraordinarily complex one with a growing literature. I have been involved and interested in it since 1968.[10] But I have never "accepted wholeheartedly," as Schaffner says I have, that selection only occurs at the gene level or denied that it ever occurs at the level of groups of individuals.[11] Instead, I am the first, last, and most explicit of the modern authors to argue (1968-78) that humans are an excellent model for group selection. They alone have the traits of conscious foresight and ability to cooperate in planning that Wynne-Edwards[12] seemed to impute to other animals in arguing for adaptive population regulation; and, because of their ability to cooperate and compete in groups, humans seem to have been their own principal selective force for a very long time.[13] I cite Lewontin in regard to levels of selection not because I am implying that he argues for the gene as the unit of selection, as Schaffner indicates, but because he states, in the paper I cite, that: "The primary focus of evolution by natural selection is the individual." Franklin and Lewontin's conclusion[14] that genes are inherited in various kinds of groups is well understood in biology, not disputable, and in no way contrary to any of my arguments. Lewontin's arguments[15] included the first discussion of precisely what determines the levels at which selection is effective in different circumstances, and they showed that significant variations in this regard are a certainty; thus, his conclusions are not accurately reflected by Schaffner's comments.

Williams's conclusion against group selection is not "a methodological assumption" but an argument now almost universally accepted in biology because of both logic and empirical evidence. Lewontin said it correctly—that population regulation, on which the argument for group selection was based, simply does not occur (at least outside humans). Other topics, like sex ratio selection, can be used to show that empirical evidence is sufficient to refute group selection in a very large realm.[16]

I end with a critical remark against myself. Since submission of my manuscript, I have continued to delve into the literature of law in relation to anthropology, sociology, and philosophy, and I am increasingly astounded and dismayed at its immensity and complexity. Anyone deeply immersed in these topics has to view my paper as naive and with numerous failures of reference. If the paper turns out to have lasting merit it can only be in regard to the approach, and the generalizations from that approach. I am thankful that I have found nothing yet which seems to me to obviate this possible virtue.

NOTES

1. See my note 11, pp. 283-84.
2. See my note 31, p. 286.

3. M. Sahlins, *The Use and Abuse of Biology* (Ann Arbor, Mich.: University of Michigan Press, 1976.

4. *American Anthropologist*, 79 (1977): 917-20.

5. See my note 14, p. 284.

6. See Fisher and Williams, cited in my notes 2 and 5, pp. 280, 281, and Alexander, in *Goulden*, cited in note 4, p. 281.

7. M. Hirshfield and D. W. Tinkle, "Natural Selection and the Evolution of Reproductive Effort." Proceedings of the National Academy of Science, 72 (1975): 2227-231.

8. R. D. Alexander and G. Borgia, "On the Origin and Basis of the Male-Female Phenomenon," in M. F. Blum, ed., *Sexual Selection in Insects*. (New York: Academic Press, 1978).

9. See also my note 9, p. 283.

10. R. D. Alexander and D. W. Tinkle, "A Comparative Review [of *On Aggression* by Konrad Lorenz and *The Territorial Imperative* by Robert Ardrey]," BioScience, 19 (1968): 245-48; R. D. Alexander and G. Borgia, "Group Selection and the Hierarchical Organization of Life," *Annual Review of Ecology and Systematics,* 9, 1978; also, all of my cited papers.

11. For example, see my note 5, p. 281.

12. V. C. Wynne-Edwards, "Self-regulating Systems in Populations of Animals," *Science,* 147 (1965): 1543-548.

13. See my note 11, pp. 283-84.

14. See Schaffner's note 4, p. 302.

15. See my note 3, pp. 280-81.

16. See my note 2, p. 280.

Part IV

Toward the Roots of
Ethics and Science

10

Science, Ethics and the Impersonal Passions

Robert C. Solomon

IN THIS CHAPTER, I would like to bring out some of the broad underlying issues governing the relation between science and ethics, the "foundations" of both science and ethics, and their common ideal of "objectivity," as expressed by the contributors to this volume. It seems to me that certain views on these matters have emerged with striking agreement, and I would like to restate them, express some misgivings about them, and then, rather than embarrassing myself by trying to sound sage in areas other scholars represented here have pursued for years, I would like to add a small contribution from some of my own recent studies.

Let me begin by anticipating the perspective I wish to pursue. Science and ethics are often discussed in terms of their results, as a set of established theories in the one case, as a code of rules called "morality" in the other. So discussed, these disciplines too easily appear as autonomous and "for their own sakes," to be analyzed but not seriously challenged or accounted for, connected to each other and to human interests and desires only contingently or even tangentially. Going back to our cultural roots, however, we can see that this intellectual and moral isolationism is a fairly recent development. As MacIntyre points out, both science and ethics are, even in the work of Isaac Newton, instances of "natural laws," God-given and purposive, the sorts of activities that things and people, respectively, *ought* to perform.[1] Going back much further, we remember that, no matter how "Platonic" the Socratic ideas, forms, and virtues, they were essentially connected to human passions. *Nous* was not just an intellectual attitude but love and enthusiasm. Aristotle would have none of our sharp distinctions between the "intellectual" and "moral" virtues and everyday emotions, moods, and desires. Where he distinguishes these, it is only, to the despair of some commentators, to reassert their unity, and his contemplative life is anything but passionless. Compare this with the Kantian views of both science and ethics, and their modern counterparts in positivism and ideal (i.e., passionless) observer theories, to see how far we have strayed from that ancient ideal.[2] It is this organic unity of science, ethics, and human passions that I want to defend, by means of the concept of what I call the "impersonal passions." It is a concept that cannot be understood as long as we think in terms of the simple Kantian dichotomy of "objectivity" and "subjectivity," with science the paradigm of the first (and thus of "rationality" too), emotions the paradigm of the latter (and thereby "irrational" as well), and ethics ping-ponged back and forth from one side to the other for the past several centuries.[3]

We can start with a paradox, a familiar one; how can science and morality be both impersonal and personally significant (i.e., motivating)? Treated as an abstract and, by definition, impersonal set of principles or rules, each gives rise to a dilemma. Why care about truth? Why be moral? Why not just be happy? It is here that desperation dictates an ad hoc answer, grown respectable only because of its urgency: "Rationality for its own sake." It is a phrase ready to the defense of any human endeavor whose significance has been forgotten. At best, it is no answer; at worst, it may be, as Toulmin says, a "pious fraud." But how is rationality motivated? What is "objectivity" that makes it our concern?

I. Science, Ethics, and "Objectivity": Some Comments

I am not going to attempt a grand Hegelian synthesis, but neither do I want to take potshots. The following remarks are some very general comments on the contributions to this volume, some of them objections that have been raised in discussion, others suggestions that have been developed in one or more of the chapters. Underlying them is an indictment of neglect. Despite repeated reference and even appeal to the "affective" dimensions of both science and ethics, these have remained vague and obscure. My comments are designed to focus attention on these neglected factors.

1. The Relation of Science to Ethics (and Vice Versa)

This grant-laden phrase confuses me. Several distinct topics have been developed. There are questions of the "external interactions" (Shaffner) or "secondary links" (Graham) between science and sociopolitical situations and ethical pronouncements, for example, the possible uses of genetic theory for selective population control or the probable uses of differential IQ tests in educational policy. There are questions of values *in* science ("first order links"), for example, the social, political, or ethical entailments of a sociological theory. The distinction between these is not always clear, and there is antagonism between those who focus on one rather than the other. For example, Loren Graham warns,

"In terms of impact, then, we are not asking the right question when we ask, 'Do scientific theories contain in themselves social or political values?' Instead, we should pay more attention to . . . the 'second order' links between science and values, those which are contingent on existing political and social situations."[4]

Stephen Toulmin has shown that many of the normative terms in science may be "ethical" only in the most superficial and insignificant way, e.g., "ideal" as used in theories of gases. Value-free science does not mean science without values, but it is not clear—particularly in the "hard" physical sciences—what sort of values might conceivably be relevant to ethics. In psychology and sociology, we can see how any theory, or even a research proposal, might affect our image of ourselves and society with ethical consequences. In biology, too, as Richard Alexander complains, even the most plausible deterministic accounts touch raw nerves—"emotional reactions" most biologists call them—and are rejected for moral reasons not usually well articulated. But, when we ask about "the relation of science to ethics," I want to be clear whether it is the ethical status of theories themselves that is in question—in which case physics and chemistry do not seem to come into it—or the *effects* of certain theories when accepted by a society, e.g., intelligence testing or genetics, or the sociopolitical parameters of the *acceptance* of certain theories by a society, or ethical values inherent in the practice of science, or the values of scientists qua scientists (which still may be very different from the values inherent in science: Stent), or the *uses* of science for ethics. And, more generally, there are dizzying moves from the relation of particular theories to particular ethical concerns to the relation of "science to ethics" as such. Individual chapters make what seem to be very significant contributions to one or another of these different questions, but the questions themselves have to be sorted out.

Another confusion concerns the nature of these supposed "relations"; it is generally agreed, and well demonstrated in Graham's chapter, that strict logical connections are too strong to characterize the "second order links" between a scientific theory and the sociopolitical climate in which it is accepted. Yet causal accounts as such seem too contingent, ignoring the "logic" of the theories and ideologies in question. (A society might be *caused* to accept a theory just because it is all that its members have ever heard, or because they were threatened with prison or divine retribution if they failed to accept it.) Some of the phrases repeatedly used in the contributions only hide the problem, for example, "fits in" or "naturally follows." It is clear, particularly in the essays by Graham and Toulmin but also in comments by Gunther Stent, Alexander, and others, that many of the parameters in the acceptance of a theory are not themselves scientific, but neither are they ethical nor ideological in the strict sense; they are what Toulmin rightly calls "affective," "personal," "emotional."[5] My complaint is that these suggestions are generally shunted aside (except by Toulmin, whom I shall be obliquely defending even while disagreeing with him). Graham once called them "nonobjective," but this is the crucial word of the collection, yet to be analyzed.

Looming over the various questions herded under the phrase,

"the relation of science to ethics," is what would seem to be the one factor shared by all sciences, whether physics or physiognomy, and ethics as well: *objectivity.* Whatever the object of science, it is to be viewed "objectively," impersonally, according to certain *rational* constraints. Whatever the situation, at least according to several dominant ethical traditions, it is to be viewed *fairly,* with a regard for *general principles,* and therefore, in an important sense, impersonally. With regard to science, we have long taken "knowledge for its own sake" to be a heroic cry, and we have looked back to Galileo with more than sympathy as the martyr to the cause.[6] Since then, we have learned that we have to question scientific investigation, even in its "pure" state, on the grounds of its technological consequences. We are just now relearning to challenge a program of research on the basis of its ethical presuppositions, something Galileo's persecutors knew well enough. For four centuries, we have defended the ideal of totally free intellectual pursuit. Now we seem ready for a swing back to human consequences as a primary concern.[7] At the end of the last century, Nietzsche asked, why must we have objectivity at any cost anyway? And here, Toulmin has adopted a distinctly Nietzschean approach, undercut the canons of scientific objectivity, and asked for the "personal engagements," the "emotional and affective components," of the scientist and his audience. In this perspective, the phrase "the relation of science to ethics" takes on new possibilities. Coupled with Alasdair MacIntyre's persistent insistence that both science and morality be viewed primarily as *practices,* we now seem to have a method not just for relating science to ethics but for asking what is their common ground, that taste for objectivity that motivates them both. And, more controversially, we can begin to ask why those practices have had such overwhelming "trump" status in human behavior. Why is knowledge superior to excitement? Morality to creativity? Objectivity to personal engagement? Rationality to romance?

As soon as we ask about the "relation of science to ethics," it becomes obvious that we have set ourselves the task of grouping together very different sets of practices into two grossly oversimplified categories. How we begin may well determine how we end up. For years, physics has been the paradigm of a science;

the Ten Commandments, the paradigm of morality. What two things could be more different? What could the "relation" between them be? Why not start, like Aristotle, with biology and questions of "natural human functions"? What happens if we consider magic? (And how would we consider it?) Why do we begin by assuming astrology is not a science but astronomy is? Or, why is concrete predictability more essential to our conception of a science than such factors as a sense of familiarity, religious oneness, and edification? Why do we think of science in terms of causal and statistical explanation rather than teleological accounts, like some of our nineteenth-century ancestors, and like some scientists today? (William Tiller of Stanford, for example, said recently that modern discoveries "suggest that at some level of the universe we're all inter-connected, as if part of one vast organism just presently becoming aware of itself.") Why not Aristotle instead of Newton, or Hegel instead of Kant? It is certainly not my intention to defend astrology or the return of science to religion, but perhaps the quest for a "relation" between science and ethics is itself the product of a false categorization and faulty paradigms.

Too often we sound as if we know full well what science and ethics are. But there are no given answers to this enterprise, since the very characterization of "science" and "ethics" is evidently a *pre*scriptive task. Whatever the characterizations, it is necessary to get beyond the usual truncated accounts in terms of principles, laws, rules, or results alone. Science is, first of all, a human activity, and to characterize an activity by its product alone to the exclusion of its process gives us the dehumanizing viewpoint in which we (as scientists) serve science as its instruments. The same is true of ethics, which is, first of all, a set of human practices, and to ignore the practice and see only its rules is, again, to reduce us to the servants and elevate the rules to rulers. (Where God supplies the rules, this might make certain sense, but insofar as we believe in autonomy in ethics in any sensible sense, this obedience model is nonsense.[8] So too in science: "the search for truth" betrays a false modesty that should, I would think, have been inappropriate since Kant.)

Science and ethics are, first of all, human activities, "practices" if not always practical. It is a matter of historical curiosity how they both got to be known as authoritative products, but it is enough for us here to point out the current curiosity. Why should we view any human activities simply in terms of products? (Perhaps sports would be a better starting point than science journals or the Bible.) I take this refocusing on practices to be the heart of Professor MacIntyre's essay, and we must add to it the Nietzschean premise of Toulmin's essay, asking *why* someone should adopt these practices. A practice *may* involve an explicit set of rules and principles (for example, the Ten Commandments and the textbook canonizations of "the scientific method"). But whether or not practices involve explicit rules, they necessarily involve motivation and personal engagement. So why have we all spent so much time looking only at the rules?

2. Intrinsic Values and Rules

In treating science and ethics primarily as practices, a recurrent point is the role of values intrinsic to those practices. Again, the problem of paradigms confronts us, because there does not seem to be any a priori reason—apart from simplemindeded thinking—for expecting the values intrinsic to psychology to be the same as those intrinsic to chemistry. But there is general agreement on at least one such value, namely, "objectivity" or "rationality." But even with this agreement, it is obvious that what counts as objective might vary among different sciences, and so might the motivation behind objectivity. It does not follow that intrinsic values are not further motivated; in fact, they must be if they are to have personal value. Here is our original paradox: how can an essentially impersonal value be a personal matter, i.e., motivating? Here, it seems to me, is one of the central disputes of the book, although surprisingly restrained. Some contributors, notably MacIntyre, emphasize "intrinsic" values virtually to the exclusion, or at least the neglect, of "external" motivation—rewards, punishments, and institutional constraints. Toulmin, on the other hand, minimizes the role of intrinsic values, calling "rationality for its own sake" a "pious fraud." Now, granted, the difference is one of emphasis, for MacIntyre fully recognizes the existence of "external" motivation and Toulmin the significance of intrinsic values. But they are pushing in opposite directions,

MacIntyre attacking the neglect of intrinsic values, and Toulmin undermining their false sense of importance. Underlying this overly quiet battle is a solid ground of agreement: both they and their two commentators (Stent and Grene) attack the idea of universal, impersonal reason, MacIntyre because it presents us with a false Kantian picture of "the autonomous individual" confronting all social relationships, practices, and institutions as "external" and therefore "positive,"[9] and Toulmin, on the other hand, because it wholly ignores the "personal engagement" of individual scientists, falsely presents the life of science as a commitment to "passionless objectivity," and consequently undervalues "the emotional or affective components in the life and work of science."[10] Both emphasize the importance of science (and ethics) as *practice,* something in which the individual becomes personally *engaged.* The difference emerges concerning the nature of the motivation for this engagement; MacIntyre focuses wholly on the intrinsic values and rules, pointing to extrinsic motives (e.g., teaching a child to play chess by offering a candy reward) only as instruments to engage him/her in the practice. Toulmin considers personal motives, including neurotic fixations and social inadequacies, in his attempt to understand the psychology of the scientist, and he treats "the rational enterprise of science" as something like one game among many which certain individuals come to engage in. What is missing from both accounts is some indication of what is special about science (and morality) and why they are not merely two practices among many. Intrinsic values are not enough. There are many practices, all with their intrinsic values, among them Alabama lynchings, garden-party etiquette,[11] and sodomy sessions, as well as science and ethics. One might speak of "for its own sake" in all of these ("one simply does not do such a thing!") but surely we want to know why science and ethics are so often said to be "higher" or more important to human nature. It is suggested that this is because of their rational character, but does this really give us anything more than a pretentious but no more revealing "for its own sake" thesis? Why should "reason" move us? Why seek truth? Why be moral? The concept of intrinsic values leaves these questions as unanswered as before. Why learn—or teach—one practice rather than another?

What I want to argue is that the distinction basic to the debate between MacIntyre and Toulmin is bogus, and that the dispute is therefore unnecessary. Of course one can identify values which *define* the scientific or moral enterprise ("intrinsic values"), and of course one can see all kinds of promises and threats used in order to get people to embark upon those enterprises ("external values"). But neither of these available findings offers us anything like an adequate account of the *personal* nature of the commitment to these objective and very special enterprises. Being moral is not like playing chess, and becoming a scientist is not like becoming addicted to acrostics. A renewed appreciation of "intrinsic values" helps us to understand what constitutes a practice, but not why one *should* enter into it. Renewed attention to the motives—perhaps suspicious or even pathetic—behind deeds or dedication may be helpful in understanding a person, particularly an unusual person, but we have yet to understand the special status of certain kinds of involvements—and certain kinds of motives.[12] What I want to suggest, therefore, is an analysis of "personal engagement" and "passionate dedication" that takes account of the efforts of MacIntyre and Toulmin but integrates them in a way that is not obvious from their own arguments.

3. The Role of the Personal

Toulmin's chapter boldly reorients us toward looking at the personal motives and engagement of the scientist. His thesis does not deny or undercut the "objectivity" of science, nor does it deny or undercut the status of science as the paradigmatically "rational" discipline. The problem, however, is that it leaves a mystery the connection between objective science and personal engagement.[13] This is nowhere more evident than in his often quoted indictment of "rationality for its own sake" as a "pious fraud." If science (and morality) are defined by their objectivity, then, on Toulmin's account, any motivation must be "external." Stent, in reply to Toulmin, gives a personal account of his own entry into science "for the good life." This is probably true in many cases, but it does not make Toulmin's thesis obvious, as Stent objects, but even more mysterious. The link between passions and objectivity is missing. And Toulmin's well-known quote from Hume, that "reason is and ought to be the slave of the passions," only makes that link even more impossible to perceive.

4. Taking Freud Seriously

Despite two excellent chapters on Freud's development and the metatheoretical significance of his models,[14] he is hardly mentioned at all in the "intrinsic"-"external" values debate. More than any author except Nietzsche, Freud ruthlessly undermined the claims of "rationality for its own sake" in precisely the way Toulmin tries to do. Consequently, he too systematically undervalued the role of "intrinsic values," but he appreciated as much as anyone the power of the "affective dimension." Furthermore, if we follow Joseph Margolis's analysis of Freud as bridging the gap between causal and personalist explanation, Freud may also offer us a way to understand the "relation of science to ethics," at least where people are concerned.[15]

5. Do Science and Ethics Need "Foundations"?

Traditionally, both science and ethics have been taken at face value, as essentially given and beyond reproach; the idea of justification, therefore, is not so much defense as analysis, laying bare the "foundations" of these enterprises. In science, suggestions range from sets of a priori principles of understanding or reason to simple sense impressions and "protocol sentences" describing them. In ethics, suggestions range from eternal ideals of rationality to biological prototypes to Divine revelation to hedonistic calculations concerning pleasure and pain. There seems to be general agreement that this old search for "foundations," from Kant to positivism and from Kant's ethics to metaethics, is ill-conceived. What might take its place, however, is not clear. Evidently, giving up the search for foundations does not amount to giving up the quest for justification. But if there are no foundations in the traditional sense, what kind of justification is possible? A cultural justification (Dworkin); a historical justification (MacIntyre); a biological justification (Alexander); a structuralist justification (Stent). But what of a *personal* justification? According to the traditional "foundations" view, this would not possibly make sense, but integrated with these other kinds of justification, including a general pragmatic theory, it can be very important indeed.

Professor MacIntyre makes the analogy between Thomas Kuhn and Kierkegaard, but what ought to be added is the sense in which Kuhn *ought* to have followed Kierkegaard but did not. Having identified mutually exclusive "paradigms" of scientific explanation, Kuhn concludes that there are *no* good reasons for choosing between them, since all such reasons (that is, scientific reasons) are within paradigms and their practices. But Kierkegaard saw that even between "incommensurable" alternatives, a different kind of justification was not only possible but necessary. (He obscured this by calling it "subjective," "arbitrary," and "irrational.") Taking "paradigms" by themselves, no comparison may seem possible. Referring rather to the life in which certain practices are adopted, one finds in Kierkegaard a personal justification for choosing impersonal principles (for example, the moral or "ethical" life). "Incommensurable" paradigms are not therefore arbitrary.[16]

6. Is It All Relative?

If science and ethics do not have "foundations," does that mean that they are merely "relative" enterprises, that their justification is never more than merely personal or cultural? Several times, this question is touched on, but rarely pursued as such. It seems to me that the general tone is antirelativist, anti-Kuhnian to be sure, but the question is never clarified and this attitude never formulated as such. (One notable exception: MacIntyre says it is clearly "wrong.")[17] When this question is discussed, it often needs to be made clear what it is that is "relative" to what. Is it beliefs, opinions, and practices that are in question, or something more, such as knowledge, truth, and right? Is it relativity to a person, a people, a culture, a historical epoch, a "conceptual framework," or human nature? This last suggestion, that something is relative to human nature—perhaps certain necessary structures of the human mind—shows that one common association, between relativism and a kind of nihilism and subjectivism, is just plain wrong. Knowledge (or whatever) may be relative to certain conditions and presuppositions, but it does not follow that there cannot be a single, ultimate set of conditions and presuppositions. Nor does it follow that different principles must follow from or have different conditions and presuppositions. Relativism leaves open the Kantian-Hegelian questions about the (human) universality of the True and the Good (and the Beautiful too).[18] Furthermore, it is often suggested that the thesis of "conceptual relativism" concerns the possibility of truly "alternative conceptual frameworks," such that there is no possibility of mutual understanding.[19] This debate, it seems to me, is a total waste of energy, even for science fiction writers. The question of relativism, if it makes any difference at all, has to work with *real* differences. But that gives us plenty to do.

As long as the question of relativism is raised with respect to the restricted notion of "conceptual frameworks," there is the very real possibility that two such "frameworks" will indeed be "incommensurable" and mutually unintelligible. But even so, it does not follow that other factors, broadly designated as "cultural" and "personal," might not make the difference. The evaluation of operational versus teleological theories of human behavior, for example, seems to me to be largely decided on the basis of extratheoretical considerations. The gaps between scientific argument and political ideology displayed in Professor Graham's discussion, for instance, may well be filled by these extratheoretical considerations. My complaint is that, except for Toulmin, these considerations are never actually considered.

7. Historical Continuity

Relativism is not, as Kuhn and Kierkegaard suggest, a matter of diachronic existential panic. Science and ethics are part of a historical development, and much of the justification of any particular theory or moral code, without being reactionary, is an appeal to historical continuity. This thesis has been challenged, of course, not only by those who refuse to recognize such considerations as justificatory but by those who deny such continuities as such, for example, most recently but with extreme obscurity, Michel Foucault.[20] Here Kierkegaard ought to have paid more attention to Hegel, for choices of existence are never made in a historical vacuum. But it is possible to go too far toward Hegelianism, and this is what I believe MacIntyre has done. It may be that our existential confusion is appropriate only to those whom MacIntyre calls "culturally marginal," but that is the lot of many of us, perhaps itself even the most prominent cultural characteristic of being brought up in America. It is here that cultural and historical considerations alone are not enough, and we must once again turn to the "personal" and the "affective." This does not mean that choices are merely individual, however. It still remains an open question whether the turn to the personal and the affective might not be at the same time a turn to essential human structures. (One immediately thinks of Rousseau, perhaps, who announced, not without pretensions, that when he looked deepest into himself, he found there what was universal to all humanity.)

8. Autonomy

Autonomy becomes important when practices begin to break down. In this sense, Kant's emphasis on autonomy is suspicious; it is not just a response to a moral crisis (although it is that too) but a defense of rationality for its own sake, which he thinks presupposes autonomy. In the Kantian sense, therefore, the concept of autonomy is more a celebration of the rational individual than it is a substantial philosophical criterion, and it is by attacking this supposed criterion that Dworkin makes a powerful case against autonomy as such as a moral value.[21] But however impressive Dworkin's arguments, I was left with the feeling that his target had ultimately been only an impoverished version of Kant, defending autonomy for its own sake, without even the pretensions of rationality at stake. What motivates "autonomists" like Sartre, however, is a very different concern. It is the irresolvable conflict of practices, principles, and "right" things to do. It is where Hegelian *Sittlichkeit* has broken down, and mere obedience to the laws of one's society becomes meaningless, since society itself contains inconsistent laws. (Hegel's obvious example is Antigone.) The same is true in science. (I here follow Dworkin's argument against Hare.) Conflicts in science are not always synchronic and not always resolvable by appeal to any current evidence, scientific practices, or cultural biases. Opposed paradigms may themselves be in question with equally commanding arguments. Autonomy is all that is left for us culturally marginal people.

It is a mistake to think, as Kant did, that rationality needs autonomy which equals human dignity. It is also a mistake to generalize the crises that demand autonomy, as Sartre does, and call it "the human condition." But it is just as mistaken to tell us that we ought to return to the primeval simplicity of *Sittlichkeit,* when such simplicity is not available, the suggestion that emerged from Dworkin, from Plato *via* Vlastos,[22] and from MacIntyre. It might be nice. But it just isn't possible. Yet autonomy must itself be investigated, to see what *personal* factors make such choices nonarbitrary.

In a short general commentary it is impossible to do justice to any of these problems or any of the theses advanced concerning them. But what I can do briefly is make a plea for a broader notion of justification and a more global stage on which to discuss these issues. What links science and ethics together, whatever their differences, is their mutual demand for objectivity, what I have elsewhere called "the transcendental pretense."[23] But science and ethics as objective practices have not yet been linked with science and ethics as personal engagements, whether motivated by the singular persuasiveness of a coherent culture or by some more individual psychological commitment which may or may not involve an explicit existential choice. I want to end this commentary by making a somewhat unusual suggestion as to what that linkage might be, concentrating on the much neglected "affective dimension" that has been so often hinted at but rarely discussed among us.

II. Objectivity and the Impersonal Passions

Even if science and morality were to be defended "for their own sakes," there would not, therefore, be any reason whatever to pursue them. But, recognizing this, it does not follow that the only motivation for adopting these illustrious practices are the crass motives of rewards and punishments, fame, fortune, parental approval, the threat of eternal damnation, and so forth. These are the twin forks of an old dilemma in ethical theory, and if Toulmin has his way they will be forks in a similar dilemma in the philosophy of science. The dilemma is usually cast in the form of a debate between Hume and Kant, on the moral role of passions, desires, and self-interest on the one hand ("inclinations") and universal laws and practical reason on the other. In our discussion, the dilemma appears in the distinction between "intrinsic" and "external" values, the "intrinsic" values, in the case of both science and ethics, falling under the rubric of "rationality for its own sake." MacIntyre emerges the Kantian, although he rightly criticizes Kant in proper Hegelian fashion for confusing the intrinsic values of a (social) practice with the autonomous virtues of reason (universal or not). Toulmin is the self-avowed Humean. Their mutual error lies in the vulgarization of the passions (enter Freud too) and the usual isolation of reason, even if both criticize it as such. What is missing is an appreciation of those passions and motives that are not in any sense directed toward self-interest but are based on universal principles. Yet they are still passions, that is, self-involved. "Rationality for its own sake" refers to just these impersonal passions. There need be no piousness or fraud. There are reasons for the pursuit of knowledge and obedience to morality that are neither crass nor empty.

What I mean by an "impersonal passion" can be explained by means of a simple example. (Hume's sympathy, by the way, would *not* be an impersonal passion on my account.) Consider the emotion, moral indignation. (The label itself should give us some clue.) Indignation concerns a sense of offense, obviously enough, but what is important is that we recognize that it is not a personal offense as such. It is a matter of *principle*. This is true even where the object of indignation is something embarrassingly petty—"it isn't the fifteen cents; it's the principle of the thing." But in indignation, as opposed to simply holding a moral view, the principle is indeed taken personally. Here is our amalgam of the personal and the impersonal. The emotion in question wouldn't be indignation unless it were based on personal principle. (Rather it would be annoyance or just peevishness.) But the principle is not something external to the emotion; it is part of the emotion itself. (Of course, it may *also* be part of an explicit moral view or theory.) And the principle is not what the emotion is about. (The emotion is about the offense in question.) The principle is the foundation of the emotional attitude itself.

As an emotion, indignation is necessarily self-involved.[24] This is a point easily abused. It does not mean that we have any self-interest in the offense which is the object of the emotion. (It might be some stranger's fifteen cents.) It does not mean that we have any self-interest in the principle. (It may even work to our disadvantage.) It does mean that we have an "emotional investment" in the principle, that we see ourselves in terms of its importance and its defense, that it is *our* principle, whether or not anyone else holds it, and even if it is universal in its form and content.

It is often thought that principles, especially objective principles, must be part of a method and have some impersonal existence of their own. Passions, on the other hand, are supposed to be strictly personal, and therefore devoid of principles (except, perhaps, certain prudential "principles," better called "habits" or "rules of thumb"). This is confusion. Thus Kant even goes so far as to argue that the support of the passions detracts from what he calls the "moral worth" of a principled action. (At one infamous point, he even argues that such passions, for example, the emotion of love, are "pathological.")[25] And science, whatever the reasons for its pursuit by a young student (Toulmin, Stent) is characterized by a method of impersonality, insured by the textbook canons of the repeatability of experiments and the total irrelevance of passionate hopes for certain outcomes, whether merely the personal ambitions of the scientist with an *emotional* investment in his/her work or broadly cultural demands (such as those discussed by Graham in his essay). But though the canons of morality and science *can* be so formulated apart from the human practices and motives behind them, they need not be and *cannot* be if we are to justify these practices in terms of these motives. The principles to be accounted for are in the motives themselves.[26] Once separated from them, whatever the advantages, all questions of motivation become either purely pragmatic and vulgar or pious and empty ("for its own sake," or "that's what it means to be rational").

Rousseau had a clear conception of the impersonal passions before Kant bastardized him in the name of practical reason.[27] Hume's often discussed notion of "sympathy" is much inferior, partly because of a general incoherence in Hume's theory of the passions.[28] The ethical impersonal passions are well characterized by Aristotle in his account of the moral virtues. (At least some of the difficulty scholars have had in accounting for the nature of the "practical syllogism" in Chapter VI of the *Ethics* is due to their loss of any such concept.) In science, the foremost modern theorist in this still unexplored area is Michael Polanyi.[29] His "intellectual passions" are precisely those passions which have impersonal principles built into their very nature. Polanyi pursues the idea, as Rousseau did in ethics, that these passions—and therefore their embodied principles—are intrinsic to human nature as such. (He even goes so far as to suggest that some such passions descend the phylogenetic ladder as far down as the worms.) It is important to distinguish, as Polanyi does not, between the impersonal passions and passion *for* something. One can have a passion for justice and a passion for chocolate, but only in the impersonal passions is a universal principle basic to the passion itself. (A passion for justice might possibly be purely personal and unprincipled, for example, in a politician in power who has a strong self-interest in seeing justice done. A passion for objectivity, for example, in a psychologist, might well be just the fear of personal involvement.)

A passionate dedication to the search for truth in science is not so much a goal as a perspective. This is why it is more than a "passion for truth," more than a motive of any kind (including those Toulmin suggests), and more than a desire extrinsic to the practice of science. This passion *defines* science, through its internal rules and demands—impersonality ("objectivity") first of all, also simplicity, a voracious concern for "the facts," an uncompromising insistence on universal agreement and proof, but mainly, a demand for familiarity and understanding on a grand scale, using a relatively restricted scheme of explanation (for example, causal explanation). It is a mistake to think that all of this is simply part of "scientific method"; if it were only that, we should have to repeat our question: why should anyone ever be motivated to take up that method? These demands are intrinsic to the passion itself. One need not (as Polanyi does) leap too

quickly to the suggestion that these are therefore "natural" out-
looks and it certainly does not follow that they are universal.
Practitioners of witchcraft are neither more nor less natural than
scientific societies. The "intellectual passions" are both genuine
passions, i.e., self-involved, and truly objective. How is this
possible? Polanyi saw the matter clearly decades ago; he gives us
a wonderful example, one often used in precisely the opposite
way:

> What is the true lesson of the Copernican revolution? Why did
> Copernicus exchange his actual terrestrial station for an imaginary
> solar standpoint? The only justification for this lay in the greater
> intellectual satisfaction he derived from the celestial panorama as
> seen from the sun instead of the earth. Copernicus gave preference
> to man's delight in abstract theory, at the price of rejecting the
> evidence of our senses, which present us with the irresistible fact
> of the sun, the moon, and the stars rising daily in the east to
> travel across the sky towards their setting in the west. In a literal
> sense, therefore, the new Copernican system was as anthropo-
> centric as the Ptolemaic view, the difference being merely that it
> preferred to satisfy a different human affection.
>
> It becomes legitimate to regard the Copernican system as more
> objective than the Ptolemaic only if we accept this very shift in
> the nature of intellectual satisfaction as the criterion of greater
> objectivity. This would imply that, of two forms of knowledge,
> we should consider as more objective that which relies to a greater
> measure on theory rather than on more immediate sensory experi-
> ence. So that, the theory being placed like a screen between our
> senses and the things of which our senses otherwise would have
> gained a more immediate impression, we would rely increasingly
> on theoretical guidance for the interpretation of our experience,
> and would correspondingly reduce the status of our raw impres-
> sions to that of dubious and possibly misleading appearances.
>
> Thus, when we claim greater objectivity for the Copernican
> theory, we do imply that its excellence is, not a matter of personal
> taste on our part, but an inherent quality deserving universal
> acceptance by rational creatures. We abandon the cruder anthropo-
> centrism of our senses—but only in favor of a more ambitious
> anthropocentrism of our reason.[30]

What I have suggested here, using Polanyi, begins with the
idea of intellectual passion in general, but it is evident from

Polanyi's example that such passions also enter into the choice of particular scientific paradigms and theories, help determine what arguments will be persuasive (Toulmin), what theories can be accepted, and what lines of research seem "worthwhile." We cannot pursue that topic here, but its outlines are clearly delineated by this notion of the impersonal passions. (It is always worth remarking that Polanyi himself, despite his seemingly more individualistic title, *Personal Knowledge,* defends a view of objectivity that is ultimately much stronger than this theory will bear.) But what Polanyi begins to give us, and what I want to suggest here, in a phrase that will make some people's skin crawl, is: *the emotional foundations of science.* From this perspective, it is not at all surprising that many of the leading terms in scientific criticism are aesthetic, for example, Feynmann on protons:

> (This) sounds simple but appears mathematically a bit unnatural. Suggestions to explain this long-range force, such as Kauffmann's, all seem a little awkward and without an inner beauty we usually expect from truth. But sometimes the truth is discovered first and the beauty or "necessity" of that truth seen only later.[31]

When these criteria are analyzed, not as metaphors but as part of the dedication to science, and when the dedication itself is seen as constitutive of scientific activity rather than a contingent and merely personal force behind it, we will be in a position to identify the missing links I have complained of. The relation of science to ethics emerges clearly, not on the level of theories and codes themselves, nor merely at the level of their consequences or "secondary links," but right at the foundations, in the impersonal passions which constitute them. As for relativism, the question then becomes an analysis of the source and nature of these passions themselves; what would it be not to have them? Could we not have them? (If not, then the usually glib claim that the demand for truth and goodness are part of "human nature" will take on some significance, and the Hegelian hope for some ultimate agreement becomes more plausible.) As for the question of "foundations" and justification, I have tried to hint that such notions must be expanded far beyond their traditional scope, to include factors usually systematically excluded. It is almost a dogma in our still too positivistic conceptions of science and ethics that the passions are not to be included in either an account or a justification of objective matters. That is, first of all, what I want to challenge. It is clear that this will involve, among other things, a reconsideration of the nature of the passions themselves, so often treated merely as visceral *cum* feeling reactions. If I am right, then emotions have a complex structure which includes much of what they supposedly motivate and defines much of what they are supposedly about. If this is so, then the "affective dimension" will become part and parcel of science itself.[32]

NOTES

1. Alasdair MacIntyre, "Objectivity in Morality and Objectivity in Science," this volume, p. 22.
2. The most recent example, though not an "ideal observer theory" in the more traditional sense, is John Rawls *Theory of Justice* (Cambridge, Mass.: Harvard University Press, 1971). The absence of personal passions is what defines the rationality of the "original position" for him (pp. 12 ff.).
3. Hume, Kant, Kierkegaard, naturalism, G.E. Moore and emotivism come most immediately to mind.
4. Loren Graham's "Eugenics and Human Heredity in Weimar Germany and Soviet Russia in the 1920s: An Examination of Science and Values," this volume, p. 140.
5. Stephen Toulmin, "The Moral Psychology of Science," this volume.
6. Camus: "Galileo, who held a scientific truth of great importance, *abjured* it with the greatest of ease as soon as it endangered his life. In a certain sense, he did right. (n. From the point of view of the relative value of truth. On the other hand, from the point of view of virile behavior, this scholar's fragility may well make us smile." *Myth of Sisyphus* (New York: Alfred A. Knopf, 1955) p. 3.
7. The extreme current version of this thesis is Theodore Roszak's *Where the Wasteland Ends* (Garden City, N.Y.: Doubleday, 1974)
8. It is clear, for example in Kant's ethics, that one can (at least try to) keep God and autonomy both.
9. MacIntyre, "Objectivity in Morality . . .," pp. 27, 28-9.
10. Toulmin, "The Moral Psychology of Science," p. 49.

11. See Philippa Foot's recent work, for example, on morality as a system of hypothetical imperatives. She makes this comparison explicitly.

12. Toulmin, "The Moral Psychology of Science," does not attempt to separate kinds of motives, and emotions and motives are not clearly distinguished.

13. Toulmin, ibid., uses the duality of "intellectually ready and psychically disposed," p. 52.

14. Steven Marcus, "The Origins of Psychoanalysis Revisited: Reflections and Consequences," this volume, and Joseph Margolis, "Reconciling Freud's *Scientific Project* and Psychoanalysis," this volume.

15. I have discussed this connection in my "Freud's Neurological Theory of Mind," in Wolheim, ed., *Freud* (Garden City, N.Y.: Doubleday, 1974) pp. 47 ff.

16. I can't help but mention that this is precisely the point MacIntyre raised against Kierkegaard many years ago in his essay, "Existentialism," in D.J. O'Connor, ed., *A Critical History of Western Philosophy* (New York: Macmillan, 1964).

17. MacIntyre, this volume, p. 27.

18. It may well be that there are concerns, structures, or parameters of human experience which are indeed common to all of us. This has been a common assumption, though defended in many different ways (e.g., Kant and Hegel) until very recently.

19. Notably, Donald Davidson, "The Very Idea of A Conceptual Scheme" *Proceedings of the American Philosophical Association*, 197; and Richard Rorty's reply, "A World Well Lost," *Journal of Philosophy*, 69, no. 19 (October 26, 1972).

20. Michel Foucault, *The Order of Things*, (New York: Vintage, 1971).

21. Gerald Dworkin, "Moral Autonomy", this volume.

22. Gregory Vlastos, "Plato's Theory of Justice," this volume, pp. 00.

23. In Meiland and Kraus, *Conceptual Relativism* (Princeton, N.J.: Princeton University Press, 1978).

24. Robert C. Solomon, *The Passions* (Garden City, N.Y.: Doubleday, 1976).

25. Kant, *Groundwork of the Metaphysics of Morals*, trans. H. Paton (New York: Harper & Row, 1964).

26. A defense of such motives in general has been worked out in detail by Thomas Nagel in his *The Possibility of Altruism* (New York and London: Oxford University Press, 1970)

27. With apologies to Ernst Cassirer; *Goethe, Kant and Rousseau*, Part II.

28. E.g., see Patrick Gardiner, "Hume's theory of the Passions" in
    D.F. Pears, ed. David Hume; *A Symposium* (New York: St. Mar-
    tin's Press, 1963) esp. pp. 38-42.
29. Michael Polanyi, *Personal Knowledge* (New York: Harper and
    Row, 1964), esp. chap. 6.
30. Polanyi, ibid., pp. 3-5.
31. From *Science,* 183 (February 15, 1974); as quoted in H. Tristram
    Engelhardt's Introduction to *The Hastings Report,* 1: 2.
32. An excellent answer to some of these concerns can be found in
    Israel Scheffler's "In Praise of the Cognitive Emotions" (*Teachers
    College Record*, December 1977, vol. 79, no. 2; pp. 171-86),
    where he deplores the grotesque split in education between "unfeel-
    ing knowledge" and "mindless arousal."

A CONCLUDING
INTERDISCIPLINARY
POSTSCRIPT

H. Tristram Engelhardt, Jr.

The series of meetings which produced this volume have, more than any of the other meetings of this series, witnessed a truly interdisciplinary examination of issues. An indication of this is found in the allusions in the papers to prior versions, earlier written commentaries, and discussions in the meetings (see, for example, Graham, pp. 30-31, Cassell, p. 20); in many ways discussions across the disciplines succeeded. The result has been a helpful interdisciplinary examination of the interplay of values and science. We have discovered that to investigate the foundations of ethics and its relationship to science is to investigate the foundations of interdisciplinarity itself. The reason should be obvious. The search for the common roots of value-theory and of science or for the intertwining of less common roots require traveling outside the confines of particular disciplines. Thus, the discussions here have been interdisciplinary: 1) by examining how issues overflow the borders of particular disciplines; 2) by determining the extent to which disciplines are not crisply delineated one from another, as is often supposed (i.e., with regard to either content or subject matter); and 3) by seeking ways of viewing the human condition which are conceptually prior to (i.e., more fundamental than) present disciplinary divisions.

There are many important conceptual problems in characterizing interdisciplinarity as Corinna Delkeskamp has shown in Volume II.[1] Should interdisciplinarity itself be a new enterprise? If so, what will bridge that new discipline and the old ones? And if it is not to be a new discipline, how is it to have integrity and adequate scholarship? One has the suspicion that in speaking of interdisciplinarity one is seeking the encompassing sense of reasoning once signaled by philosophy—a sense which embraced logic, epistemology, ethics, *and* the sciences. In any event, these three years of meetings have yielded not only a picture of how investigations do not conform to disciplinary geography, but also some notion of how to travel without the usual maps. These rules for the traveler have included such guides as the five themes suggested in the Introduction for sorting out interdependencies of content and of method between ethics and the sciences. The next and final volume of this series will address the character of the analyses of these relations in greater detail. In doing so we will be moving closer to examining the foundations of an analysis of the foundations of ethics and its relationship to science. As Solomon's commentary suggests, such a move may not involve a retreat to greater abstraction, but rather an advance to the heart of the matter.

NOTE

1. Corinna Delkeskamp, "Interdisciplinarity: A Critical Appraisal," in *Knowledge, Value and Belief*, ed. H. Tristram Engelhardt, Jr. and Daniel Callahan (Hastings-on-Hudson, N.Y: Hastings Center, 1977), pp. 324-54.

Index